SICK OFFICE SYNDROME

THE DEVASTATING EFFECTS OF LONG-TERM EMPLOYMENT

First published in Great Britain 1995
by Victor Gollancz
An imprint of the Cassell Group
Wellington House, 125 Strand, London WC2R 0BB

Text © Ian Carney 1995
Illustrations © Carl Flint 1995

A catalogue record for this book is
available from the British Library.

ISBN 0 575 06149 9

Designed and typeset by
Fishtail Design

Printed and bound in Great Britain
by The Guernsey Press Co. Ltd, Guernsey, Channel Isles

SICK OFFICE SYNDROME

THE DEVASTATING EFFECTS OF LONG-TERM EMPLOYMENT

BY
IAN CARNEY

WITH ILLUSTRATIONS BY
CARL FLINT

VICTOR GOLLANCZ
LONDON

For Carole and Jake
I. C.

Special thanks from the heart of my office to
Pete Bishop
Woodrow Phoenix
Ian Harrison
Chris Webster
and Ed Hillyer
C. F.

Introduction

A lot has been made by press and politicians of the problems and pressures that face the long-term unemployed. It is claimed that those without a job for long periods of time suffer from the effects of routine, boredom and lack of purpose, often leading to depression, crime, suicide, drugs . . . tragically, even daytime television.

Close inspection, however, reveals that the unemployed don't have it too bad.

The hours are pretty good.

You're your own boss. The only real decision you have to make is whether to get up before lunch or tea.

The money isn't great, but let's look at this in detail. The average unemployee receives £91.00 benefit every fortnight. To earn this all he has to do is visit the local DHSS centre and sign on.

This takes what? Half an hour? £91.00 for thirty minutes' work. That's a payrate of £182.00 an hour, putting the jobless up in the supertax bracket. If the Government would only tax them accordingly then the country could shift its unemployment problem over to offshore tax shelters.

Let's face it – it's the long-term *employed* who have a hard time.

Sick Office Syndrome

Sick Office Syndrome (SOS) is the nightmare that social commentators are really afraid to talk about. Those with office jobs suffer from routine, boredom and lack of purpose, often leading to depression, crime, suicide, drugs . . . tragically, even Spud-U-Like lunches. The syndrome generally appears a couple of years into an individual's working life, and takes the form of extreme or obsessive behaviour as a way to block out the reality of a sad nine-to-five existence.

The nature of this obsession stems from a facet of the employee's behaviour that is already there. If, as a child, he particularly enjoyed stamping on ants, chances are that after five years of working in an office he will be capable of planet-wide genocide. If she had a secret childhood language of her own creation, she'll probably end up as a company spokesperson, unable to talk anything but fluent officialese bullshit.

Any ambition or sense of achievement is quickly destroyed as the years of low reward/minimal responsibility fly past.

If, by some fortuitous conjunction of circumstances, a sufferer *does* earn a promotion, then the symptoms will be carried up to a senior, more intensive level (which explains why your boss has the mental flexibility of a hatstand).

Sick Office Syndrome is a malaise that strikes only *office* workers. Manufacturing and retail work could also be considered repetitious workplace-fixed professions but certain factors prevent these workers from suffering from SOS in its purest form.

All retail staff have the options of being rude, unhelpful and helping themselves from the till, so they really don't deserve any sympathy (with the notable exception of the fine shop staff who stock this book, who are doing a magnificent job).

By the time this book comes out the country won't have a manufacturing industry so their employees can also be ruled out.

No, it's the office workers who sit and suffer – and all work and no pay desk jobs make Jack/Jacqueline/Jock a dull/compulsive/dangerous person.

Remember

Having a job for life is not something to be ashamed of. It could happen to anyone. It probably happened to you.

Preparing for Long-Term Employment

An office job never fails to come as a shock.

You stumble through childhood with dreams of becoming a cowboy or the first woman on the moon and then . . . BAM!

It's forty years sitting in an enclosed workstation pressing FUNCTION 1.

Maybe children should be eased into their working lives with toys such as Data-Processing Barbie or Action Manager. How about a transformer that changes from an exuberant youngster into a washed-out, unlikeable worker drone? As it is, there's only the computer game LEMMINGS (© Psygnosis Ltd) that gives a child the sense of what it's really like 'out there' – endlessly trooping alongside exact duplicates of yourself, with nothing save suicide on your minds.

Schools could add WORK TEDIUM to the syllabus – double periods of pretending to look busy, making personal phone calls and displaying scarcely disguised contempt towards middle management.

A Job for Life

If your job has survived eight terms of Conservative rule (accurately predicted to give this book maximum shelf-life) then chances are it's a job for life.

You may as well face the fact now that in years to come your grandchildren will sit on your knee and ask to hear of the great moments in your life. Those times of action and intrigue, when the mighty flow of history hinged on an action you made or a decision you took. All you'll have to tell them about is that time when you helped unblock the photocopier and the agonizing choice between requisitioning blue or yellow Post-its.

Electronic Age

Don't go pinning your hopes on the advent of the electronic office combating SOS, improving working conditions and changing the very structure of the working day.

As those already out of the computer honeymoon period will vouch, all the electronic age will ultimately do is replace the rows of worker drones sitting at desks shuffling papers overseen by an oppressive office manager with rows of worker drones sitting at workstations shuffling data overseen by an oppressive systems manager.

The Varied Strains of SOS

What follows is a catalogue of the most common strains that Sick Office Syndrome takes, how they affect the sufferer and the long-term prognosis of their influence.

Everyone currently working in your department will be suffering (albeit some only lightly) from at least one of the strains, and some of the more hopeless cases will be victims of a combination of two or three. *Don't be surprised if you find yourself exhibiting the symptoms of all of them.*

The Surly Yoof

The Surly Yoof will enter the office and immediately suffer, her young body being unable to put up any resistance to SOS. Shyness and inexperience will be tossed aside in favour of teenage arrogance. Every order she's given will be met by tutting and a rolling of the eyes. She will turn up her nose-ring at any task that takes longer than her twelve-second MTV attention span.

Being born in the fast lane of the Information Super-highway (instead of languishing in the bus lane like the rest of the staff), the Surly Yoof could make a real difference to the smooth running of the Electronic Office. E-mail and Macs will be as familiar to her as quill pens are to the senior management, but when asked for her assistance she'll just say she can't be arsed.

To her, everything is SAD. When pressed to name the good things in life she'll mumble out some soundbites about the glorious punk revolution, 1977 and idiot dancing to Lurkers b-sides. Basic mathematics will reveal that she wasn't actually born until 1978.

The Surly Yoof is the one member of staff who'll be pleased with her salary, the measly income being triple the amount she got in her last job (a paper round). Unfortunately, it won't stretch as far as a bottle of Clearasil.

Visual Characteristics
Shaved head, post-E culture fashion-victim clothes and cider-stained DMs. She will possess so many facial piercings you could strain spaghetti through her cheeks.

Diet
Jolt cola and eight bags of Quavers (which provide the body's daily requirement of monosodium glutamate).

Role Model
Tank Girl's uncommunicative younger sister.

Favourite Music
The next big thing. If Raygun predicts a streetwise Sousa revival she'll be out there looking for the Tricky dubmix of 'Washington Post'.

Habitat
With her parents in a six-bedroomed detached house in the suburbs. It's somewhere for her to hang her CLASS WAR t-shirts.

Prognosis of Prolonged SOS

Her sullen behaviour ensures she never lasts beyond her probationary period.

The New Lothario

A sexual predator from an early age, this self-proclaimed 'shag-meister' actively pursues an office career because of 'all the boss skirt that works there'. Soon afterwards he'll realize the terrible mistake he has made. Even so, his constant arousal will have the effect of diminishing the blood flow to his brain, causing slight mental stupefaction, ideal for long afternoons data controlling.

Like all good hunters, the New Lothario has had to evolve his stalking techniques to keep track with behavioural changes in his prey. During the eighties, he boasted of his monthly condom bill and spouted implausible tales of sexual exploits with minor porn stars, double-jointed gymnasts, Siamese twins, etc. Unfortunately, the 'If I said you had a beautiful body would you let me shag you senseless?' approach no longer cuts it in these post-feminist/PC times. Even the greenest temp can sniff out this bullshit (no matter how it's disguised with *Dolce and Gabbana Pour Homme*).

Now he has evolved a New Man approach, reading poetry in his lunchbreak, crying at pictures of puppies and espousing feminist manifestos.

The female staff soon fall over themselves to lend this sensitive flower a shoulder to cry on – the shoulder being only 'a hop, skip and a Wonderbra away from the tits!'

NB The major difference between the Lothario and the Lech is that the former actually gets shagged occasionally.

Visual Manifestations of SOS

Once sub-George Michael, now a delicate sub-Daniel Day Lewis (albeit Daniel Day Lewis on a bad hair, face and body day).

Diet

'Frankly I find it hard to indulge, not when there's so many starving people in the world. When I think of those hungry children ... You'll have to excuse me, I've got something in my eye.'

Habitat

Will redecorate his flat, removing the mirrors on the ceiling/Pamela Anderson calendar/notched bedhead features and replacing them with CND posters/Jeanette Winterson novels/pink cushions to show he's in touch with his feminine side. The fact is, if he did have a feminine side he'd be so in touch with it he'd be all over it like a rash.

Catchphrase

Once, 'As long as I've got a face, you've got somewhere to sit, darling!' Now, 'Is it just me or is there something about whalesong that moves you to tears?'

Marital Status

If this manifestation occurs in an older employee, chances are he'll be married with children, but he'll not let this deter potential victims. His wife will be used as a sympathetic lure ('She can't seem to understand my need to just *be*') and the kids as further proof of his sensitivity ('I cut their umbilical cords, all the time sobbing with joy!')

Prognosis of Prolonged SOS

Over the years will adopt a variety of disguises to ensure successful 'sniffing', even briefly returning to his gelled-hair, 'naturally' tanned, medallion-man image when the retro-eighties look comes into fashion.

The Gambling Man

This man's SOS symptoms stem from the basic idea that one hefty bet on a rank outsider will be his passport to an employment-free lifestyle. After a successful win on the office Grand National sweep this seems temptingly possible, and every minor success extends his gambling behaviour further.

Not content with throwing his money away on horses, dogs, the pools, fruit machines, the National Lottery, charity scratch cards and multiple tabloids for the bingo inserts, he'll also initiate bets on pointless day-to-day activities in the office. He offers good odds on how many times a day the photocopier will break down (7–4), how many personal phone calls the new office temp will make (11–2) and the possibility of aliens emerging in the Admin department and devouring the company's VAT records for the 1986–87 period (12,000–1).

The Gambling Man also develops a series of strange behavioural patterns which he believes bring him luck. If he receives a static shock from the office carpet on the morning of a successful win, then he'll believe this is lucky and will be seen violently rubbing the floor on the morning of every bet. Before long he'll have to place a hat on the telex machine, say hello to someone called Pam, cross and uncross his legs 37 times, receive a static shock from the office carpet and crash into a parked police car before he can even think of going to the bookies.

Visual Manifestations of SOS

A galloping speech pattern that is somewhere between John McCririck and Red Rum. His extravagant hand movements can either be put down to tic-tac mannerisms or a nervous twitch at the thought of his life savings being placed on Crispy Lawrence in the 2.35 Back-ease Support Bra Novices Hurdle at Towcester.

Habitat

Spends long hours on the phone to RAPID RACELINE (39p a minute cheap rate), finding out the latest from Kempton, Chepstow and Haydock. Wastepaper bin piled high with discarded betting slips.

Marital Status

Likely to settle down with a telephonist whom he'd describe as a 'well regarded, useful filly who's not an easy ride but who finishes well.'

Office Affiliations

Is the key member of the office National Lottery syndicate. Unfortunately, he'll recruit so many members that a maximum payout would earn them only £18.37 each. He'll particularly bond with the Lech as they both share an interest in watching small men performing on animals.

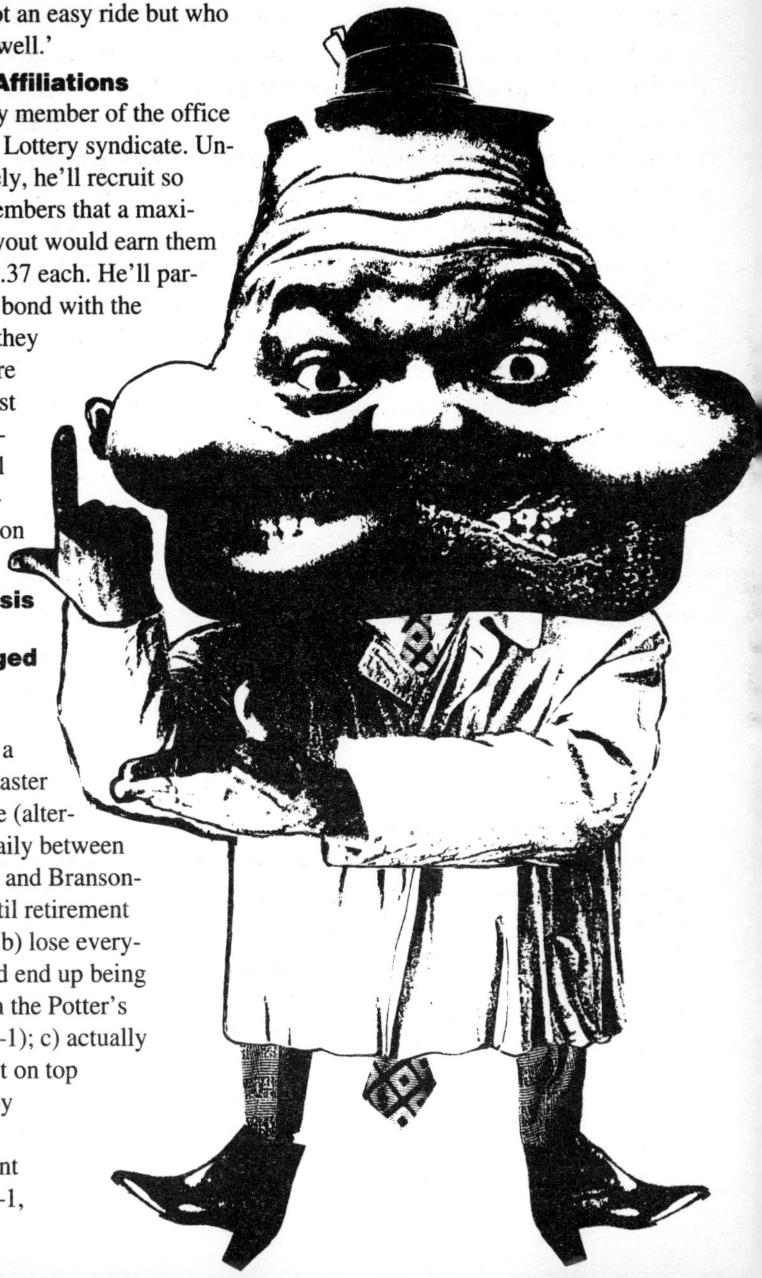

Prognosis of Prolonged SOS

He'll a) carry on a roller-coaster existence (alternating daily between destitute and Bransonpeer) until retirement (evens); b) lose everything and end up being buried in the Potter's Field (2–1); c) actually come out on top and enjoy an early retirement (10,000–1, sucker!)

Thatcher's Child

Weren't the eighties marvellous? This Young Conservative thought so, believing every bit of the 'future is golden' spiel and borrowing, borrowing, borrowing in order to spend, spend, spend.

His job was something vague in the City and his satellite-dish-sporting *des res* was packed to the mahogany dados with matt black flatpack furniture; a CD player in every room and an XR7 on the driveway.

Then the tide turned. His job evaporated, the flatpack furniture fell apart and his house was repossessed. But no quality time on the dole for this chap. Instead, he was forced to take a crappy office job to appease his creditors, secretly hoping to ride the next mini-boom out of their clutches.

Unfortunately, the boom never came and SOS has begun to bite, causing Thatcher's Child to live in his gilded past to avoid his grey present. To him, every day is 1984 and a fake-tanned, Filofax clutching YUPPIE is a thing of great beauty and not just something for the younger generation to mentally file alongside QUAKER and MINER as archaic images of times gone by.

Remember: Thatcher's Child is the spawn of the devil and deserves no sympathy. Feel free to twang his braces/steal his executive toys/spit in his Perrier.

Visual Manifestations of SOS
Sharp, expensive suits which look increasingly worse for wear. Often sits at his workstation with a far-away look on his face, mentally reliving paintball weekends past.

Habitat
The Poll Tax and Pony Tail, an eighties theme pub decorated in attractive investment portfolios.

Diet
In his mind it's Dom Perignon and sushi; in reality it'll be super-market-brand cola and Pot Noodles.

Most Treasured Possession
A 'You don't have to be a grasping megalomaniac to work here . . . but it helps' plaque, a dog-eared souvenir of better times.

Prognosis of Prolonged SOS

Will work, work, work to earn, earn, earn the money to pay back the amount he borrowed, borrowed, borrowed to spend, spend, spend. Unfortunately, because of the sum involved and his current pitiful income, he won't be out of the red until 3010.

The Serial Redundant

The Serial Redundant will have difficulty in staying in a job for more than a couple of months. Not through any fault of his own, you understand, he'll just enter the firm as it hits a downward turn or as the management's streamlining initiative comes into effect, and as he'll be the last one in, he'll be the first one out. Again.

He will immediately be shunted into another desk job which will last even less time.

After this occurs half a dozen times or so, he'll start to suffer from accumulative SOS, actually expecting redundancy from the moment he completes an application form for a prospective job. He'll walk any interview he sits (by this time he'll have an impressive CV of work experience) but starts his new position a quivering wreck, dry heaving every time he's handed a brown envelope or summoned to the Managing Director's office.

The fact is, the Serial Redundant won't actually *need* to work as over the years he'll accrue a fortune in redundancy pay-offs, a nest egg that he'll be frightened to touch 'because of the uncertain job climate'.

To spot the Serial Redundant in your office, just mention the phrases HOSTILE TAKEOVER or MARKET TESTING and he will identify himself by hyperventilating.

Visual Manifestations of SOS

A nervous twitch. Will insist on carrying around his P45/references. May even keep his coat on as he's 'probably not stopping'.

Habitat

Nomadic, moving to wherever the work is.

Will never enjoy a desk of his own (no point making yourself comfortable if you're going to be clearing out your drawers in a week or two) or become a home owner (obtaining a mortgage would be impossible with his spotted career history).

Greatest Achievement

Will accumulate Europe's finest collection of cancelled ID badges.

Effect the Electronic Age will have on Relieving His SOS

Will actually intensify his symptoms by encouraging his (not misguided) fear of being replaced by a computer.

Office Affiliations

Will obviously never make any lasting ties during the fleeting visits he makes to various offices.

His one chance to secure some like-minded friends will come when he helps establish a Redundancy Support Group for fellow victims. Unfortunately, before he can bond with any of his fellow members, he'll be dismissed during a membership down-scaling.

Prognosis of Prolonged SOS

Will finish working his way through every office in Great Britain just as the European workplace truly opens up.

The Habitual Post Graduate

Years of hard study and financial sacrifice have brought the Habitual Post Graduate a first in something useful and relevant such as *Egyptology*. Suddenly forced back into the real world, she has discovered that there aren't many pyramids excavated in downtown Lewisham.

She will be forced to take an office job 'just for a year', to pay off her student loan/credit card/student union bar tab.

Unfortunately, she will then be snagged by SOS, her ambitions drained and a regular income making escape into the arms of further education seem impossible.

And so the Habitual Post Graduate desperately clings to her university lifestyle, living in a state of denial of the nine-to-five present and taking every opportunity to impress her collegiate background and superior intellect upon her workmates.

To make matters worse, a classmate who left school at sixteen without any qualifications will end up as her line manager.

Visual Manifestations of SOS

She will dress ex-student drab, not even a monthly wage being able to drag her away from the charity shops. Will often be seen in the office canteen reading the latest pop 'difficult' novel.

Musical Tastes

Will continue to 'wig out' to Joy Division and The Smiths, believing them to be top alternative fodder and not realizing that they are to the 'yoof of today' what Yes and Barclay James Harvest were to her generation.

Habitat

A bedsit, with her real-ale-supping, Internet-addicted, long-term boyfriend who claims to be a *mature* student and yet acts like a twelve year old. Their first child will be christened *Ramses Ptolemy Theakstone's Old Peculiar Jones*.

Personality Quirk

An annoying habit of writing messages in hieroglyphics when sending Christmas cards, which she considers an individualistic touch and the rest of the office thinks of as pretentious shite.

Prognosis of Prolonged SOS

Will get to use her archaeological prowess on reaching retirement, when she examines a life in ruins.

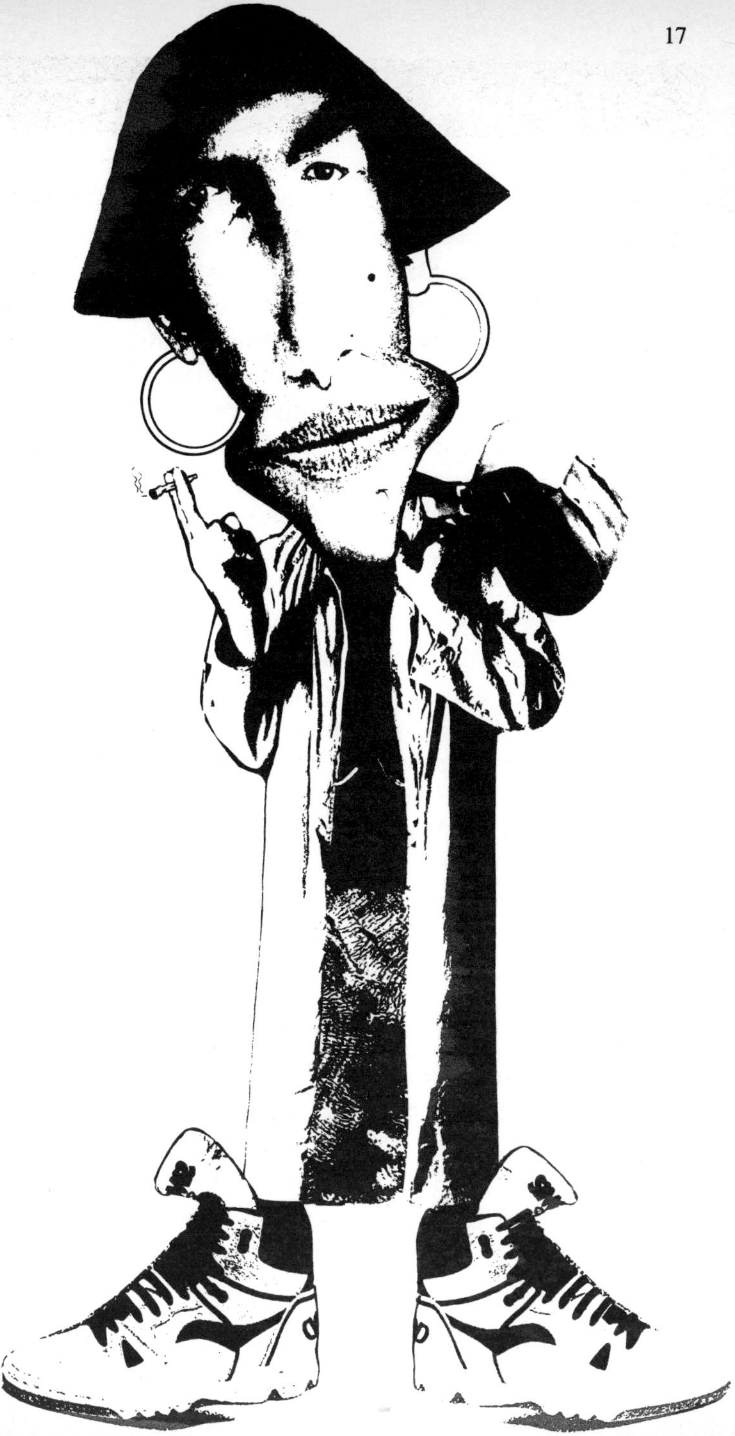

The Piss Artist

Late, hung-over morning arrivals, long liquid lunches and repeated early darts of an evening mean that the Piss Artist is constantly struggling to make up his flexi-time. It also means that he's constantly struggling to walk straight, stay awake and not puke in his out-tray.

His arrival at this behaviour pattern occurs after making a simple mathematical deduction:

SOBER = conscious of the boredom of work.

DRUNK/STONED = anaesthetized to the boredom of work.

Because of this, he takes every opportunity to drink himself numb, his wages being paid straight to the brewery by direct debit.

In moments of real desperation, when opening time is whole minutes away, he will drink the white-spirit-based VDU cleanser, followed by a Tipp-Ex chaser.

Talking to him is like conversational Russian roulette: depending on when you catch him he'll either be your best friend in the whole world or he'll want to punch your head in.

NB To ensure your survival, never attempt to drink in rounds with him.

Visual Manifestations of SOS
Red eyes, red nose and, after fifteen years of alcohol abuse, red urine. Outer clothes marked by telltale curry/vomit stains.

Social Skills
Can burp/fart the theme to *Raw-Hide*.

Chemical Dependencies
Only indulges in prescription drugs. Unfortunately, they're prescribed by Mad Barry from Data Support Group. The Piss Artist is eternally grateful that the office's no smoking policy doesn't extend as far as crack cocaine.

Greatest Achievement
The yard of ale followed by the two trouser legs of urine.

Effect the Electronic Age will have on Relieving His SOS
Very little as he won't be able to focus on the VDU on his desk. Believes E-mail is drugs sent through the post.

Habitat
Any licensed premises within stumbling distance of a curry house.

**Prognosis of
Prolonged SOS**
His future looks hazy.
Then again, so does
his present.

The Scary Weirdo

The Scary Weirdo is a career psychotic. He will start his working life strange and SOS ensures that by retirement he will be completely batshit (or he'll be transferred to Rainhill, whichever comes sooner). He is so creepy that there'll probably be an X-file on him.

Whether talking to himself in a voice so strangled that he makes Mr Bean sound like Orson Welles, or initiating letter writing campaigns to the BBC (as *Points of View* proves, only scary weirdos write letters to TV companies), he is living proof that Care in the Community doesn't work.

He will develop a disturbing obsession with true crime, collecting *World of Serial Killers* (a 107-part weekly series – free blood-splattered binder with the first issue) and planning alternative routes on a road map of Hungerford. If you comment that it's a nice day, he will reply that the murder rate increases in direct proportion to rises in the temperature. If you mention the frost on the ground, he will explain how frigid conditions delay body decomposition.

NB If you ever mock the Scary Weirdo, you will undoubtedly be included in his little red book and so had better become accustomed to the underside of patios.

Visual Manifestations of SOS

Intense, staring eyes and extended incisors. Will often be seen carrying a torso-sized sportsbag.

Habitat

Will live at home with his much-mentioned/never-seen mother. More frighteningly, he'll KNOW WHERE *YOU* LIVE.

Effect the Electronic Age will have on Relieving His SOS

Uses the office word processing facilities to upscale his letter writing campaigns.

Strange but True

The Scary Weirdo's personality matches every murderer's psychological profile in the first five series of *Crimewatch*.

Prognosis of Prolonged SOS

What, you haven't seen *Silence of the Lambs*?

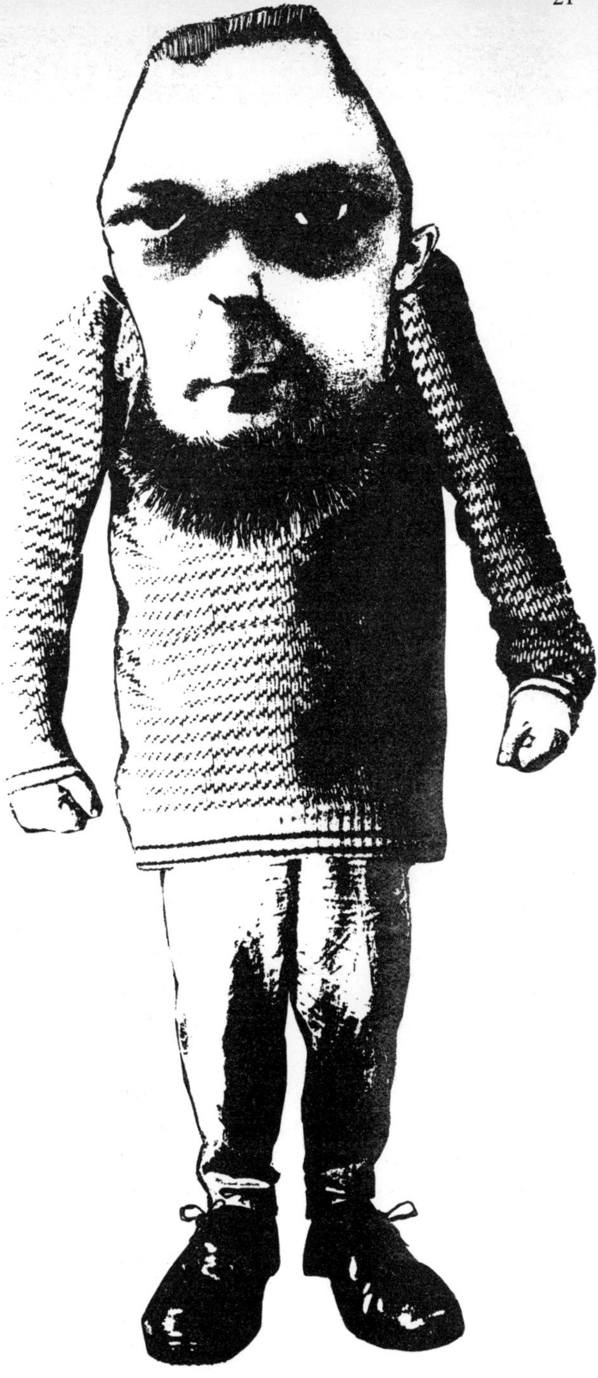

The Anal Retentive

It won't be difficult to spot this sufferer's workstation. Papers stacked to attention, in-tray constantly empty and pens methodically arranged in an industrial-sized desk tidier, the Anal Retentive's naturally neat and systematic nature will be spruced by SOS into a full-on, hands-washed-thirty-eight-times-a-day, colon-irrigating obsession.

Not believing there is a limit to the amount of time one can daily brush one's teeth/alphabetize one's files/verify one's stationery allocation, the Anal Retentive will right-angle around the SOS 'ward' exuding efficiency.

Unfortunately, he'll spend so much time keeping his desk 'ship-shape and Bristol fashion' that he won't actually get any work done.

Crease a page in his Roladex and he'll weep like a baby.

Visual Manifestations of SOS

AT WORK: hair mathematically parted, trousers pressed to attention, shoes pristine. RELAXING: hair mathematically parted, jeans pressed to attention, trainers pristine.

Habitat

A spick and span bachelor pad which he'll vacuum twenty-three times a day. Move an ornament, even a millimetre, and HE WILL KNOW.

Diet

Will enjoy square meals and rectangular snacks and will think crinkle cut chips an abomination. There is a noticeable absence of roughage in his diet.

Perfect Mate

Someone clean and angular with a fresh pine aroma who cleans around the bend after flushing. Ideally, a toilet duck.

Prognosis of Prolonged SOS

Will frottage himself to death in an unfortunate buffing incident.

The Working Mum

It all seemed so easy. She'd have the baby, enjoy a relaxed maternity leave, lodge the kid with a child-minder, head back to work and carry on where she left off. Hey, Shirley Conran said life-juggling was a breeze!

Except it didn't work out that way. She did not reckon on SOS magnifying all her anxieties and making the duties of motherhood seem far beyond her. To make matters worse, the baby senses its mother's distress and displays (secondary) SOS symptoms of its own.

So the Working Mum works an eight-hour day after no sleep (the baby teething/suffering from wind/scared of the dark) and all her clothes are formula/shit/vomit-stained and she won't have time to iron them anyway and she'll use all her annual leave until 2014 when the baby gets chicken pox and her flexi-time will be three years in debit from long lunches shopping for nappies/Calpol/Milton and her designer living room will look like an assault course worthy of *The Krypton Factor* and she won't be able to keep her mind on the job because she'll be worried about the baby and will have to phone the child-minder every ten minutes to tell her to hold a mirror to her mouth to check she's still breathing (the baby that is, not the child-minder) and she won't have time to *think*, never mind have sex . . .

Visual Manifestations of SOS
See above. A mumbling wreck clutching an economy pack of Pampers and a Mothercare carrier bag.

Office Affiliations
Will immediately bond with any woman who's given birth. Together they'll be able to clear a room in seconds with talk of loose green stools, piles like bunches of grapes, stretch marks, etc.

Diet
No time to eat. Can be seen munching on rusks if desperate.

Effect the Electronic Age will have on Relieving Her SOS
None. She will, however, blame VDU radiation cast during her pregnancy for mutating her baby into the crying/pissing/puking monster it is today.

Prognosis of Prolonged SOS

Symptoms of SOS will actually begin to fade as child reaches school age. Life will just begin to settle into some semblance of normality when baby number two comes along.

Senior Management Slacker

This character joined the company straight from school. Which means that the person running the office has suffered forty years of SOS, making his derangement the most pronounced in the place.

After so long at the workface, his sickness takes the form of work aversion, dodging responsibilities and bunking off like a schoolboy. He will continually stress that the secret to total quality management is *delegation* which, as everybody knows, means getting everybody else to do the work.

The bulk of this will fall on his Personal Assistant (see Utilitarian PA) whom he treats in a chummy and rather condescending manner. He will never comment on the sterling work she does, but occasionally throws in a criticism just to 'keep her in her place'.

The only time he pitches in is when the company is expecting a visit from head office/shareholders, and then he's so diligent the place appears to revolve around him.

Detached, immobile, selfish and forgetful, it is quite obvious that there is a thin but definite link between senior management and senile dementia.

Visual Manifestations of SOS

Waddles along the corridors like a pin-striped walrus (see Diet), huffing through cigar smoke-encrusted nasal hair and loudly greeting anyone he passes (though he'll have no idea who they are).

Diet

Exists on five-course power lunches with other Senior Management Slackers, a clandestine world of Glenfiddich quaffing, Portillo praising and Masonic handshaking.

Habitat

A well positioned office of gargantuan proportions (to practise his putting) with an en suite toilet (somewhere to sit and read his Scottish golfing holiday brochures).

Office Affiliations

The only time he mingles with other staff members is at retirement parties, where he provides a set speech (written by his PA with INSERT NAME HERE at the top and a photograph attached so he can recognize the employee concerned).

Effect the Electronic Age will have on Relieving His SOS

He can page Scottish golfing holiday brochures on the Net.

Prognosis of Prolonged SOS

Eventually retires and therefore experiences absolutely no noticeable change in his daily routine.

The Utilitarian PA

As the Senior Management Slacker's . . . *slacking* intensifies, and he encourages his Personal Assistant to make more decisions and handle his workload, she will eventually become the real power base in the office. Displaying intelligence, diligence, confidence and lots of other 'ence' qualities, she will eventually possess everything that *he* lacks.

The day-to-day running of the company will fall to her, the big boss merely signing whatever paperwork she passes across his desk. She could probably issue a directive authorizing the use of unregistered firearms to deal with awkward clients and he'd happily add his ornate scrawl to the bottom.

She also runs his personal life, sending flowers on his wife's birthday, renewing his annual golf club subscription and arranging his dinner dates. If she only had time to be insulting to shop assistants, dribble gravy down his suit and make love to his mistress once a week then her boss could stay at home in bed.

Things will begin to get a little sticky during the PA's annual leave. The phone rings off the hook, nothing gets sent out and her office looks like a paper chase, as the temp brought in to replace her gets stuck into this week's *Hello!*

The PA will return the following Monday, twitch her nose like Samantha in *Bewitched*, and the place will be back on track in minutes.

Not much sick about the Utilitarian PA, you may say, but what you have to remember is that she does *all* this work for a teeny-weeny Personal Assistant's salary.

Sounds pretty sick to me.

Visual Manifestations of SOS

Neat and efficient, any time you see her she'll be simultaneously typing a memo, answering the phone and making the tea.

Undoubtedly has the ability to pat her head and rub her stomach at the same time.

Habitat

A tiny office adjoining her boss's room, which despite its crampedness will contain all his work.

Diet

Never seen to eat or drink or even go home at night. Close monitoring reveals that she sneaks a few chocolate Hob-nobs when preparing tea for a client (Iceland Jaffa cakes when it's no one important).

Catchphrase

'He's in a meeting at the moment. Can I take a message?'

Prognosis of Prolonged SOS

Will replace her boss in a reverse Mrs Doubtfire manoeuvre.

The Lager Loutette

Quiet, intelligent and demure on her own, this girl turns into a foul-mouthed, lager-swilling, sexist yobbo when mixing with a group of like-minded drinkers, SOS causing them to adopt a hive mentality as they rush to deaden brains reeling from their dreary existence.

The collective will congregate to celebrate paydays, birthdays, days of the week with y in them and hen nights (the prospective bride recognizable by the GAGGIN' FOR A SQUADDIE Post-it stuck to her back). No man will be able to walk past them without receiving a cry of 'Get your todger out for the girls!' or, 'There's not much of that to the kilo!' followed by gales of staccato laughter.

Any man who reacts adversely to these advances will immediately be branded a miserable bastard or gay.

Nights out will be a blur of debauchery – dissected in infinite detail (on company time) until the next night out. Take any of the girls away from the group and the Lager Loutette symptoms immediately fade, the subject reverting to her original personality before adopting one of the other SOS personas.

Visual Manifestations of SOS
See the 'JUST IN' rail of any high street outlet, complemented by warpaint worthy of *Sitting Bull*.

Habitat
Anywhere there's male strippers or a late licence, preferably somewhere with both.

Topics of Conversation
Brad Pitt's buttocks, Ryan Giggs' stamina and Daley Thompson's lunchbox.

Diet
Slim-Fast all week until Saturday night when it's twelve glasses of lager, five brandy and Babychams, four packets of crisps, a five-course Indian Banquet and a kebab on the way home.

Lager Loutette's Ideal Man
A nubile young thing with a stomach as flat as an ironing board.

Lager Loutette's Actual Man
A lethargic couch potato with a stomach as flat as an ironing basket.

Prognosis of Prolonged SOS

Faces either cirrhosis of the liver or Demi Moore/Kirk Douglas *Disclosure*-style sexual harassment suit (but without the dodgy virtual reality bits).

The Brown Nose, Face and Header

Serving as nursery milk monitor, going on to be teacher's pet in infant school and then becoming a 'fag' to the older boys in senior school, it is not surprising that this worker will develop an eagerness to please equal to the Circassion harem slaves of Suleiman the Second.

He will settle into a central position in the office, living to serve senior management and generally displaying more suck than VAX VERSUS THE GIANT MUTANT ANTEATER. His days will be spent faxing, delivering shopping for senior management, and his lunchtimes providing an excellent car valet service for them.

Unfortunately, despite all this, he will constantly be overlooked for promotion – after all, he is too busy servicing his superiors actually to do his *own* work, and if he became management grade, who would send faxes, deliver, shop, etc?

The Brown Nose, Face and Header's accommodating nature will extend only as far as senior employees. He is indifferent to anyone in his own grade and downright cruel to anyone below him, believing himself deserving of the same level of ass-kissing that he delivers himself.

Visual Manifestations of SOS

Snivelling, rodent-like features. Some more advanced cases develop the word WELCOME across the forehead.

Habitat

Prostrate at the feet of his direct line manager. Will eventually wear a groove running from his desk to the sandwich shop and back to his boss's desk.

Favourite Little Job

He will constantly instigate collections for managers' birthdays and marriages. Even WE JUST WANTED TO SAY WE LOVED YOU, BOSS collections, using a combination of threats and blackmail to ensure a maximum payout from the underlings. Because of this, any whip-round for the Brown Nose will always end up in debt.

Office Affiliations

Will form a disturbing sub/dom relationship with the Middle Management Sadist.

Prognosis of Prolonged SOS

Will enjoy a post-office life career as a Tory backbencher.

The Despoilt Hire Plant

The only staff member who is actually a vegetable (although some of the upper grades could be considered border-line), the Hire Plant will be installed in the office by an outside contractor to provide an attractive working environment. This is, in fact, the only type of plant hire currently taking place in non-industrious Britain.

Seeded in a cloistered nursery and then cultivated in a temperate greenhouse, the Plant will have difficulty surviving in the inhospitable terrain of the modern office. Choked by cigarette smoke/BO/re-conditioned air; deprived of photosynthesis-generating sunlight (strip lighting just doesn't cut it); leaves either plucked and rolled like some kind of executive toy, smoked as a narcotic or taken as a cutting; roots poisoned by discarded coffee dregs, fag ends or vomit/urine (after particularly intense lunchtime sessions), it has to be *bloody* hardy even to survive its first day.

The Hire Plant is attended to once a fortnight by a mal-odorous little man with green fingers (not to mention a green neck, green hair and a green butt-crack).

Visual Manifestations of SOS

Starts off as a vigorous ornamental specimen with a sturdy stem and stunning foliage and ends up as a stick with a stunning leaf (and to be honest, it's not that stunning).

Diet

Will develop a strong chemical dependency, eventually escalating into a £15-a-year Baby Bio habit.

Habitat

An ornate pot containing dense, slow-draining topsoil and an alkaline subsoil.

The soil level will increase at the same proportion as the employees' escape tunnel proceeds (see Middle Management Sadist).

Greatest Fear

That staff members have taken so many illicit cuttings it'll receive a visit from the child support agency.

Claim to Fame

A distant relative was once blanched on national television by Sophie Grigson.

Prognosis of Prolonged
SOS

Will be replaced by an
identical specimen,
in plastic.

Mister Muscles

As a child this staff member developed a healthy interest in exercise after reading an advert in an American comic. Depicted in crude pictorial strip, the ad told the story of a teenager humiliated in front of his girlfriend when he gets sand kicked in his face by a musclebound bully. The teenager develops a set of muscles, punches out the bad guy and becomes 'Hero of the Beach'. Impressed by this simple morality tale, our employee has worked out at every opportunity, intent on building himself a body just like Charles Atlas (and not meaning dead for years).

He takes an office job only to supplement his gym fees/vitamin input/Lycra collection but soon feels the drag of long-term employment. After a particularly lengthy early-morning workout he discovers that the endorphins in his bloodstream help numb the pain of office life so effectively that he increases his exercise schedule to a masochistic level. Soon his diary reads like the first week of the Olympics (gym, swimming, running, tennis, squash, parading around in a blazer to the national anthem) and his body swells to funhouse mirror proportions, breasts growing larger than any of the girls' in the office.

NB What Mister Muscles doesn't realize is that in the caring, sharing nineties, the Hero of the Beach is no longer a teenager who cracks jaws but a pale skinny thing who cries, writes poetry and tells his best friend he loves him.

Visual Manifestations of SOS

Even fully clothed he still looks like a cross between a shaved gorilla and an over-inflated bouncy castle.

Diet

A high-energy protein supplement that provides enough energy to run eight marathons a day. He'll consume this and then attempt to sit still at a terminal for eight hours.

Habitat

At desk, clenching and unclenching his buttock muscles in order to expend some of his energy (see above), which gives him the most highly developed sphincter in the inner city area.

Spends every lunchbreak at the gym and every afternoon smelling like he has.

Illegal Habit

Develops a taste for black market steroid cocktails that, despite the fact they cause hair loss, secondary female characteristics, aggressive behaviour and impotence, increase his chest by 0.34cm.

Prognosis of Prolonged SOS

As his body grows to Canary Wharf-size proportions, his head will disappear altogether in a scene reminiscent of the closing sequence of *The Incredible Shrinking Man*.

Mister Mid-Life Crisis

This staff member is one of the lucky few who can work for years without showing any noticeable SOS symptoms. Indeed, he actually embraces the routine, allowing the repetition to numb him until he eventually returns home to his Women's Guild wife and his collection of antique Dinky cars.

But then, as retirement draws uncomfortably close, he will suddenly look along the recurrent corridor of his life and find it leads directly to . . . DEATH! If only he could turn back time . . .

Almost overnight he transforms. His comfortable bifocals are replaced by contacts! A lemming-like toupé! Primary-coloured suits! A children's television presenter's personality! Assertive sexuality! His wife will look on bemused as he clears aside his Gerry Rafferty CDs to make room for the latest electro compilations, soak himself with colognes such as Testos for Men and L'Homme Rampant, and trades in his four-door saloon for a two-door 'pussy magnet' (well, next door's cat will sleep on the bonnet).

He'll attempt to thwart the Grim Reaper with a formidable routine of squash, windsurfing, club attending and leopard skin pouch posing, not realizing that this sudden strain on his system will actually serve as a GRIM REAPER! THERE'S A STIFF ON ITS WAY OVER HERE neon sign.

Mister Mid-Life Crisis will honestly believe that the young people in the office see him as 'Experienced But Still Happening. Respect!' In actuality, they will see him as 'Comedy Old Person'.

Visual Manifestations of SOS
See above. Imagine your Uncle Eric in a reversed baseball cap.

Habitat
Perched on the corner of female employees' desks, Barney Rubble socks exposed to confirm he's a *fun guy*. Nights will be spent in those bars where people pretend they're in Ibiza and not Birkenhead.

Favourite Holiday Location
Mini-breaks in Torquay will be replaced by Alaskan Snow-

boarding holidays and marlin fishing off the Bahamas. As a bonus, he is probably eligible for subsidized flights for the elderly.

Most Embarrassing Display of SOS

His answerphone message, which he'll *rap* over a suitably 'fly' backing track.

Prognosis of Prolonged SOS

Symptoms will end as abruptly as they started after sufferer glimpses his 'new self' in a shop window and realizes that death is preferable to such indignity.

The Rumour Mill

Any move you ever make, any conversation you ever have with a member of the opposite sex, any telephone call you ever receive, will all be noted by the Rumour Mill. The facts will then be churned through her sick mind until they emerge as either:

a) a story of your dishonesty

or

b) a story of your perversity

So, when the rumours circulate about you being found naked in a petting zoo, you'll know where they started. Same place as the story about you selling company secrets to the Welsh. Being naturally curious, it is understandable that SOS will affect her in this way and, let's face it, percolating malicious gossip is much more fun than filing any day. SOS will also cause her to develop a memory longer than an elephant on smart drugs. Any tiny indiscretion that you make will be forever stored in her mind, every embarrassing detail recorded in digital clarity.

NB Any true *act of office debauchery will send her into a seizure.*

Visual Manifestations of SOS

Big eyes – all the better for seeing.
Big ears – all the better for hearing.
Big mouth – all the better for spitting out venomous half-truths.

Habitat

Wherever there's loose lips. Often found lurking around other staff members' phones and pressing the last redial button or dialling 1471 to find out who's talking to who. What who is *doing* to who will then be supplied by her imagination.

Weakness

Acts with fever-pitch indignation if any rumours circulate about her.

Greatest Regret

That she can't steam open and read other people's voice mail.

Effect the Electronic Age will have on Relieving Her SOS

Little, but she will establish an office rumours bulletin board on the Internet.

Prognosis of Prolonged SOS

Just between us, she will be found in a compromising position with an accounts clerk, a German Shepherd (not the species of dog) and two litres of strawberry yoghurt (low fat).

But you didn't hear that from me.

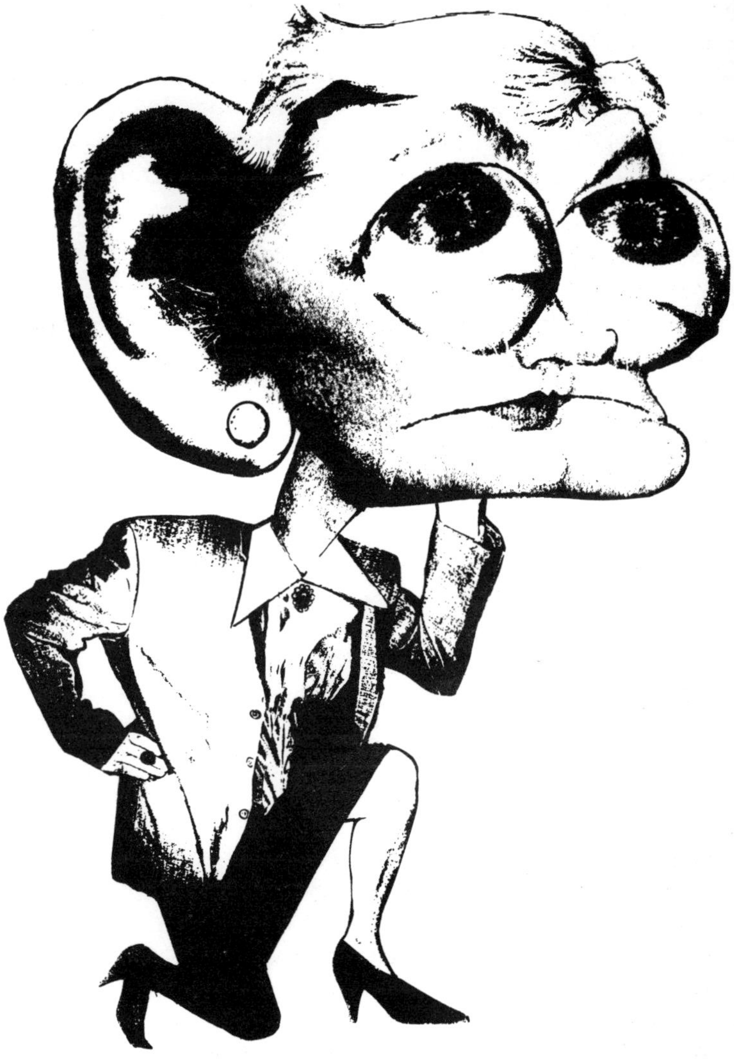

The Lech

After a hard day's work, this fairly normal-looking employee (hands a little sweaty, trousers a little crusty) will return to the privacy of his back bedroom and become a full-on, hard-loving, loin-flooding sex bison . . . by himself.

He'll start his career a shy young thing who finds it difficult to communicate with the opposite sex. Finding that real-life sexual politics don't run as freely as they do in the letters in *Forum*, he will gradually withdraw from humdrum office life and sink deeper and deeper into a world where girls are called Roxy and always have nymphomaniac twin sisters with a penchant for fresh fruit.

SOS will further tilt his mind until he finds very little difference between reality and the images in his low-resolution, fifteenth-generation videos, reaching for the PAUSE button every time he catches a glimpse of bra strap. Girls will easily be able to identify the Lech by his accommodating hands and the way he spends conversations addressing their sternums.

Visual Manifestations of SOS

Bulging X-ray eyes, hands thrust into split-crotch pockets, neck cricked upwards from lunch hours scanning the top shelf.

Right arm developed like a Gladiator.

Ideal Woman

A busty, wasp-waisted blonde with black tape across her eyes and insufficient funds to pay the TV repairman.

Greatest Achievement

Over the years will pick up a working grasp of the Swedish language.

Habitat

Positions his desk near the toilets so he can nip for some quick 're-lief' should any female staff members actually talk to him.

Prognosis of Prolonged SOS

He'll discover the twenty-four-hour porn channel and starve to death.

The Sick Man of Europe

Early on in his career, this employee will make the simple deduction that illness brings him much-needed time away from the office. SOS will cultivate this discovery into a raging obsession, with more and more obscure diseases claimed, some of a tropical nature rarely found in this hemisphere.

The Big Book of Non-Specific Viruses will become like a bible to him, one particularly wild sickness claim securing him the cover of *The Lancet* and causing germ warfare experts to quarantine the entire office building.

The Sick Man of Europe will be off so often that you probably wouldn't recognize him if you bumped into him. In fact if you did bump into him he'd probably stay off sick for six months with some kind of vague internal injury.

Visual Manifestations of SOS

Will make a point of always being seen clutching a rainbow-coloured tissue to his nose whilst coughing in a volcanic fashion, dramatically foreshadowing a prolonged bout of sick leave.

Diet

Claims to have difficulty keeping anything down – 'Did you know that pâté comes back up looking like walnuts?'

Most Ludicrous Illness Claim Used on a Self-Certification Form

Repetitive strain injury from completing self-certification forms.

Most Disgusting Illness Claim Used on a Self-Certification Form

A toss-up between 'anal weeping' and 'flaky sack'.

Effect the Electronic Age will have on Relieving His SOS

Computers will have little effect in relieving his psychosis but they will give him a whole new palette of illnesses to 'suffer', including work-related upper-limb disorders, stress, backaches, non-ionizing radiation poisoning and a sore foot (a disk drive fell on it).

Greatest Regret

That he was born male and can't claim any of the ever-reliable, non-challengeable 'woman's problems'.

Prognosis of Prolonged SOS

Will live in perfect health until he's one hundred and three.

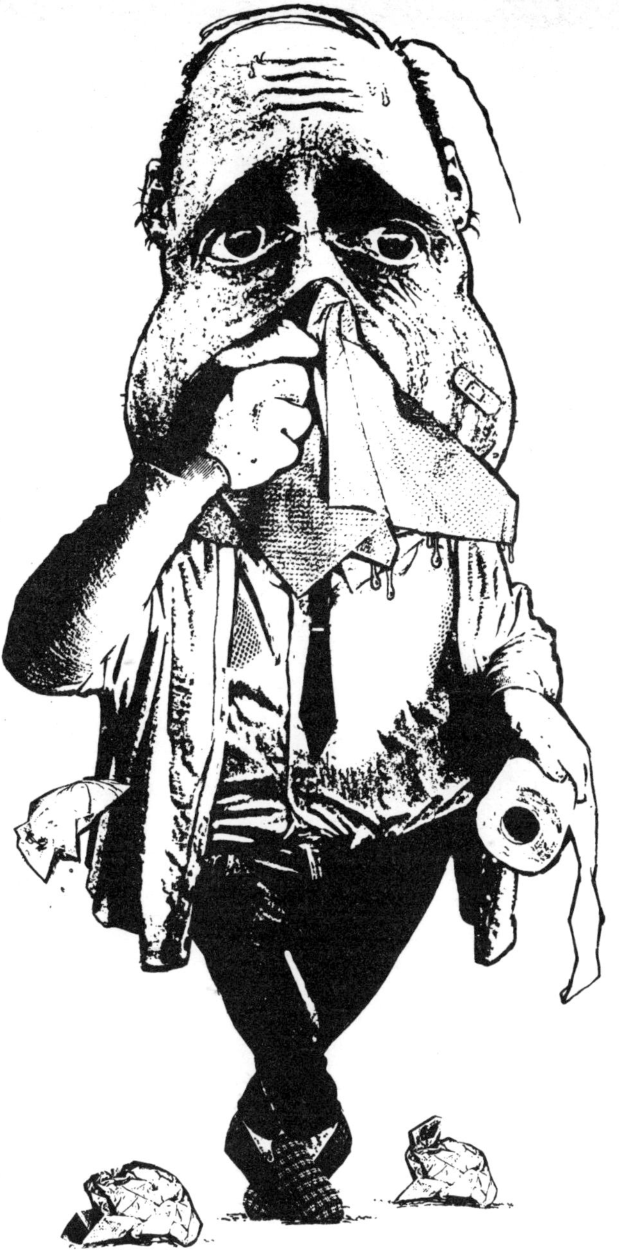

The Office Clown

This serious young man takes an office job as a temporary stop-gap until he can gain a foothold in the undertaking industry. After a relatively short time, SOS will bring his comic 'genius' to the surface as he becomes a devotee of the 'You've Got to Laugh or Else You'd Cry' school of thinking, who believe that the only escape from the daily grind is with a laugh and a smile.

Whether it's faxing photocopies of his genitals to Head Office or attempting the 'Can-I-use-your-dictaphone?-No-use-your-finger-like-everyone-else' joke, the Office Clown will constantly try to provoke hilarity.

Unfortunately, he'll generally fail.

If, by some fluke, one of his inane remarks does manage to raise a titter, he repeats it endlessly in the mistaken belief that repetition makes it even funnier.

NB This is not the person to share an enclosed train carriage with on long journeys to off-site training seminars. Not unless you've got a good barrister.

Visual Manifestations of SOS

An irritating smirk and a vast array of cornea-bruising ties. The first sign of warm weather and he'll don a pair of tailored, knee-length shorts.

Habitat

Anywhere he's tolerated. Laugh at one of his jokes and you've got a friend until retirement.

Social Skills

Generally considered the life and arsehole of the party. Will karaoke at the drop of a hat (favourites include 'My Ding-a-Ling', 'Donald Where's Your Trousers?' and Side 1 of *Dark Side of the Moon*).

Greatest Accolade

The company Sports and Social Committee will unanimously vote him 'Employee most likely to be beaten to death at 2 a.m. in a pub car park'.

Mating Call

A wide repertoire of catchphrases purloined from a variety of phrases including 'Hu-Uh-Hu' (Beavis and Butt-Head), 'What's the Frequency Kenneth?' (R.E.M.) and 'That's a marlin, not a suppository, Lawrence!' (source unknown).

Prognosis of Prolonged SOS

He will marry young, his wife realizing her mistake on the wedding night. Their subsequently messy divorce will cause him to hit the bottle, get sacked and later develop cancer. 'Still, you've got to laugh!'

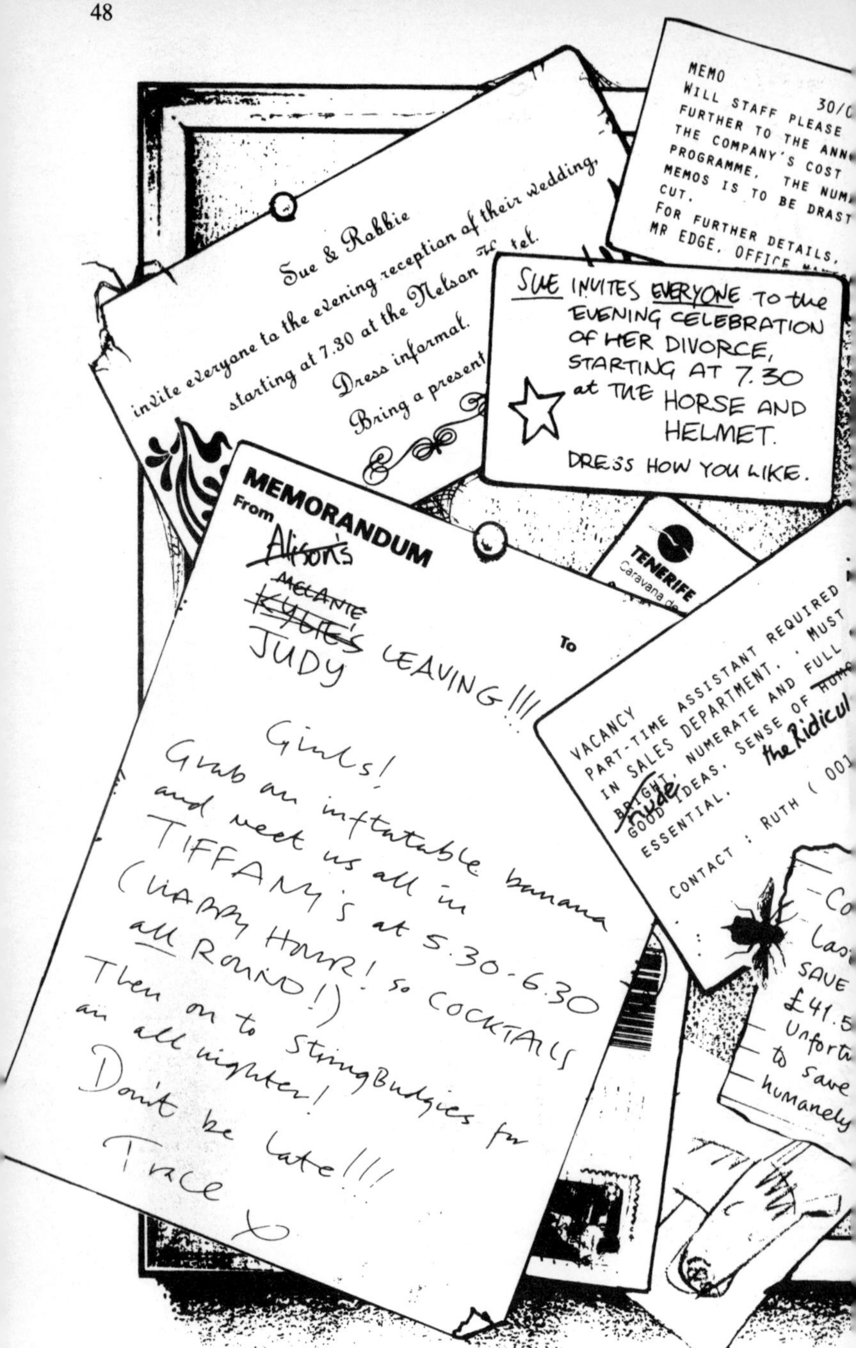
MEMO
30/0
WILL STAFF PLEASE
FURTHER TO THE ANN
THE COMPANY'S COST
PROGRAMME, THE NUM
MEMOS IS TO BE DRAST
CUT.
FOR FURTHER DETAILS,
MR EDGE, OFFICE
Sue & Robbie
invite everyone to the evening reception of their wedding,
starting at 7.30 at the Nelson Hotel.
Dress informal.
Bring a present
SUE INVITES EVERYONE TO the EVENING CELEBRATION OF HER DIVORCE, STARTING AT 7.30 at THE HORSE AND HELMET.
DRESS HOW YOU LIKE.
TENERIFE
Caravana de
MEMORANDUM
From
~~Alison's~~
~~MELANIE~~
~~KYLIE'S~~
JUDY
LEAVING!!!
To
Girls!
Grab an inflatable banana and meet us all in TIFFANY'S at 5.30-6.30 (HAPPY HOUR! so COCKTAILS all ROUND!)
Then on to StringBudgies for an all nighter!
Don't be late!!!
Trace x
VACANCY
PART-TIME ASSISTANT REQUIRED
IN SALES DEPARTMENT. MUST
~~BRIGHT~~ Rude, NUMERATE AND FULL
GOOD IDEAS. SENSE OF ~~HUMO~~ the Ridicul
ESSENTIAL.
CONTACT : RUTH (00
Co
Las
SAVE
£41.5
Unfortu
to save
humanely

FIRE ALARM INSTRUCTIONS

As it took well over an hour more than the acceptable time limit to evacuate the building last fire drill, will staff kindly observe the following !

In the event of a fire, staff should not:

1) Sit around ignoring the ringing bell for at least ten minutes, thinking its either a car alarm or the return of their tinnitus.

2) Spend precious, life-saving minutes collecting together personal items of value.

3) Slowly meander down to the stairway discussing last night's Eastenders.

4) Ignore any attempt to register their safe departure from the building by nipping off to the pub.

5) Use the excitement as an excuse not to do any work for the rest of the afternoon.

6) If anyone knows the whereabouts of the Piss Artist (missing since last drill), please contact personnel.

OFFICE NETBALL TEAM

Girls ! New members required to inject spirit into flagging team. Members must supply their own white t-shirt/ gym slip/ Reebok Instapump Vert HXLSs.

Sign name below (office lech need not apply).

CAN'T 1 EVAN BE THE
SPONGE BOY ?

MEMO   03/05/97

WILL STAFF PLEASE REFER TO MY MEMO OF 30/04/97 AND REMEMBER TO ENSURE THAT THE NUMBER OF MEMOS ISSUED IS DRASTICALLY CUT.

FOR FURTHER DETAILS, CONTACT MR EDGE, OFFICE MANAGER

UNION MEETING !

Will all members congregate in the Union Room at 12.30 to discuss the management's derisory pay offer. We shall be expecting a free and frank discussion to establish an acceptable offer based on realistic terms well in line with inflation.

Geoff, if you can't make it I'll meet you in the pub after work.

Colin

PS Can you lend us a fiver

The New Ager

Starting her career as a down-to-earth pragmatic type, SOS has caused the New Ager to seek an escape from reality with various Eastern religions and philosophies. She has worked her way through *Buddhism*, *Hinduism*, *Jainism*, *Sikhism*, *Parseeism* and *Confucianism*, in fact anything of an Oriental origin.

She was halfway through David Ickeism before she realized he was born in Leicester and not Nepal.

The New Ager fervently embraces the concept of karma, coming to the conclusion that she must have been Attila the Hun in a past life to deserve the nine-to-five torment she currently suffers.

When not at work she seeks spiritual validation in the local alternative, counter-culture, smash the system, anti-corporate bullshit 'head' shop (Access, Switch, Am-Ex accepted).

Visual Manifestations of SOS

Her personal aura is 'as beautiful as a Painted Lady butterfly emerging from its cocoon on a glorious summer day'. Which makes up for her looking like a sad hippy who's never owned an iron.

Social Life

Won't get out much because she can't find a suitable baby sitter for her inner child.

Diet

A strict vegan (she won't eat *anything* with a brother-in-law), her diet consists entirely of pulses. Which explains why she looks like she hasn't got one.

Habitat

Insists her crystal-laden desk is situated with strict regard to Feng-shui and ends up sitting adjacent to the third urinal from the left in the male toilets.

Prognosis for Prolonged SOS

Will come back in the next life as someone in the same job, except with longer hours and less pay.

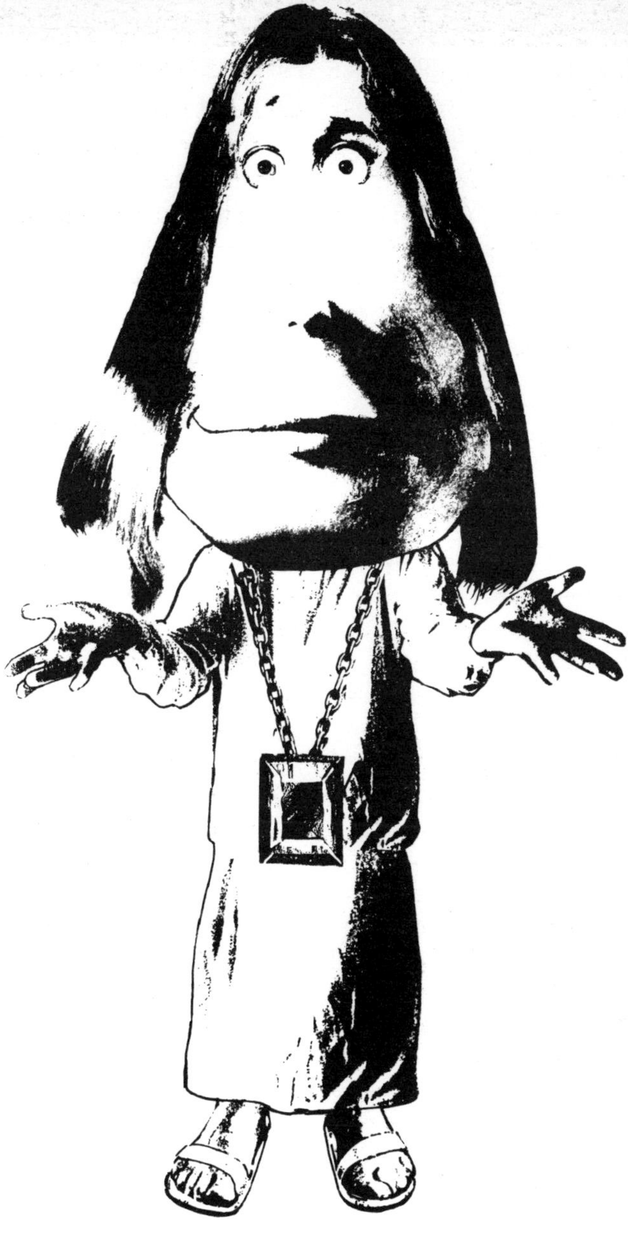

The Courier Out of Hell

A childhood visit to see Eddie Kidd open the local abattoir made an indelible impression on the Courier Out of Hell. The biker lifestyle offers a multitude of attractions – the oily beards, the bell-hipped biker girls, the crusty leather jackets that contain so many badges that they resemble armadillos . . . No being manacled to a terminal in an over-air-conditioned cell for him! Instead he'll be on the open road! The wind whistling around his speeding form! Insects hari-kari-ing into his open mouth! Long months having steel pins inserted into joints after life-threatening crashes!

A motorcycle courier will seem the ideal job – he'll get paid to be in hog heaven. Unfortunately, long hours waiting in office reception areas (often spending more time in the place than the 40 h.w. desk jockeys) and breathing the same air as crazed long-term employees will ensure that the Courier Out of Hell will suffer Passive SOS.

More and more he will sink into his Loner of the Highway persona, the world through his eyes resembling a post-apocalyptic wasteland, humanity's only chance being the package that he must deliver to the solicitor's across town. He will drive increasingly fast, take more risks and gain more replacement limbs/organs as his career flourishes.

Visual Manifestations of SOS
Matt-black helmet with face visor and a padded suit, giving him a fearsome Judge Dredd-like appearance. Start worrying when he begins to pack an MK15.

Habitat
His day will be split between the highway (5%), bleeding in a layby (15%) and waiting around/completing paperwork (80%).

Favourite Holiday Destination
An annual pilgrimage to Douglas Central Hospital, Isle of Man.

Greatest Achievement
Breaking Evel Knievel's long-standing record. Not the one for jumping over buses, but the record for the amount of facial skin grafts someone can receive before their face is officially an arse.

Prognosis of Prolonged SOS

Will be forced to take a job in the mail room after one crash too many

pushes his insurance premium into six figures. Will replace his RIDE FAST, DIE YOUNG t-shirt with a CRASH A LOT, SPEND A LIFE-TIME PUSH-ING A PARCEL TROLLEY comfy sweater.

The Vamp

It's a power thing.

Entering the office intent on using her intellect to carry her to the top, this over-achiever will spend the first few years of her career working long and hard for little reward.

Then one day she'll come to work in a tight t-shirt and be amazed at the change in her male colleague's attitudes towards her. In a sick office you have to play dirty, and so her subdued business suits will be replaced by outfits more clingy than a cling-film salesman's handshake; her bookish demeanour shelved in favour of a psychosexual assertiveness that has her male superiors sitting up and begging. Sexuality is a powerful weapon in the hands of a post-feminist manipulator.

Almost overnight, the adoption of a little girl voice delivering lines like, 'I always find stockings so much more comfortable than tights,' will cause her work caseload to halve, boosting her power player position into the premiere league.

NB The Vamp would never dream of shagging any of her 'suitors', but the possibility that she might will be enough to keep them interested. If any of them ever tried anything she'd probably break their arms.

Visual Manifestations of SOS

Come-to-bed eyes, kiss-me-deeply lips and a tan-me-with-a-lightly-greased-spatula body, the Vamp will look like every man's dream plaything but is nobody's fool.

Diet

Whatever she wants from wherever she wants. All she'll have to do is mention it out loud and there'll be someone who 'just happens to be passing there on their lunchbreak'.

Catchphrase

'Is that a Series 3a Psion organizer in your pocket or are you just pleased to see me?'

Habitat

Can be found either sitting at a desk positioned so her male co-workers can see her legs or in the daydreams of the aforementioned male co-workers.

Prognosis of Prolonged SOS

Will use her intelligence and sexuality to take her to the very top. Hopefully, things will then change . . .

Mister Teflon

Lubricated by SOS, Mister Teflon will slide through his career avoiding any confrontation whatsoever. As slippery as an eel in baby oil, he will continually manage to wriggle out of any trouble (usually of his own making) and come up smelling of roses.

If the company invested heavily in GIZMO production just as the bottom fell out of the GIZMO market, a deal that originated on *his* desk, then Mister Teflon would profess ignorance and adopt a pious attitude towards 'whoever' thought GIZMOS were happening in the first place. Of course he'll have signed his name to nothing. If, however, the GIZMO investment hits paydirt then *he* was the wheeler-dealer who landed that sucker – hey, GIZMO is his middle name.

More untouchable than Elliot Ness, he will develop a sixth sense for trouble, going out for lunch just seconds before a snarling phone call from a dissatisfied client, leaving one of Mister Teflon's colleagues to deal with it.

Try to confront him and you'll find he's *just* nipped out for a couple of minutes.

No matter how hard the company is hit by recession and overseas competition, you can be assured that the shit will never hit Mister Teflon's fan. In fact it will be a non-stick fan, trained in clandestine Ninja shit-avoiding techniques.

Visual Manifestations of SOS

An ethereal, almost-presence about the office, only glimpsed out of the corner of your eye as he leaves the room.

Habitat

Never here, always there.

Catchphrase

'Nothing to do with me, mate.'

Office Affiliations

Will never really bond as people will have difficulty getting to know the 'real' him, especially as he has more faces than the *Swatch* catalogue.

Prognosis of Prolonged SOS

Will eventually become so accomplished at his subterfuge that he'll manage to avoid coming to work altogether and yet still remain on the payroll, receiving a substantial annual promotion and subsequent payrise.

The Computer Bore

This is the technophile who is paged every time your disk drive isn't driving, your visual display unit isn't displaying or you're cursing your cursor.

Equipped with components foil-wrapped like Pop Tarts, he'll fiddle around the back of your terminal for half an hour, all the time attempting to impress you with High Level Language lifted straight from *Star Trek*: ('That's the problem with your RS-232 – zero output cripples your online flow!').

Concluding his visits by drawing attention to your VDU's NO FOOD OR DRINK TO BE CONSUMED NEAR KEYBOARD sign (which you'll follow by bringing his attention to your hand-written DON'T CONFUSE ME WITH SOMEONE WHO GIVES A FUCK sign), the Computer Bore is friendly, efficient and obviously deeply immersed in every aspect of his job.

This man is in the ideal career – something he enjoys doing and that he approaches with great enthusiasm. Not a hint of sickness here.

However, a happy madman is still a madman. He'll love the job because he genuinely considers himself to be a futurist at the vanguard of the computer age, proud to be at the cutting edge of the user/system interaction wedge. In reality he is just a glorified TV repairman with a dress sense that borders on the criminal.

File under DELUSIONAL.

Visual Characteristics

An anonymous little man with a user-unfriendly personality and an interface only his mother could love. Surprisingly, considering his tidy, linear mind, he'll dress in a ramshackle fashion, not even a modem being able to connect his shirt tails to the back of his pants.

Character Quirk

Actually buys the computer magazines that everyone else just reads in W.H. Smith's.

Diet

He'll claim it's '*chips* with everything!' This isn't true. On Thursdays he has boiled potatoes.

Effect the Electronic Age will have on Relieving His SOS

The electronic age will be the basis of his sickness, every second at the circuit board helping to compound it.

Computers will become his life

and he'll think of everything in electronic terms, going as far as describing his performance in bed as HARD DRIVE. Unfortunately, his wife rates him somewhere between SOFTWARE and FLOPPY DISK.

Prognosis of Prolonged SOS
He will be replaced by a computer.

The Middle Management Sadist

Absolute power may corrupt absolutely, but the Middle Management Sadist is proof positive that a modicum of power also corrupts . . . and brings out a criminally cruel streak.

He will emerge from the rank and file, the chance of promotion seeming like a hand up from the clutches of SOS.

Unfortunately, he will soon find that all he has done is risen to a higher, executive level of SOS, where the added pressure encourages him to run his department in a military fashion. The particular military he has in mind being the *Beheime Staatspolizei*.

Every minutiae of the staff handbook will be enforced with a regime of intimidation, humiliation and pain. Show any signs of weakness or nervousness and he'll be on you like a ton of bricks, questioning your every PRINT function and visit to the toilet.

His ruthlessness will extend to himself, giving himself a damn good beating every time he's a 'naughty boy' (about four or five times a day).

Visual Manifestations of SOS

Steely eyes, military bearing and a walk that just falls short of a goose-step, he believes that career wear would be greatly complemented by a leather mask with a zip across the mouth.

Habitat

Secures a desk near the exit which he'll man like a guntower. Don't be surprised if employees attempt to tunnel their way out.

Biggest Disappointment

That he couldn't persuade Head Office that an unmerciful spanking was just reward for continual bad timekeeping.

Favourite Reading Matter

Anything on personnel management or something light by the Marquis de Sade.

Catchphrases

No pain, no gain/You have to be cruel to be kind/Time for the nipple clamps, PAYE dolt!

Prognosis of Prolonged SOS

He will eventually be harassed out of his job by Amnesty and then immediately replaced by an exact duplicate.

Meet the new Middle Management Sadist, same as the old Middle Management Sadist.

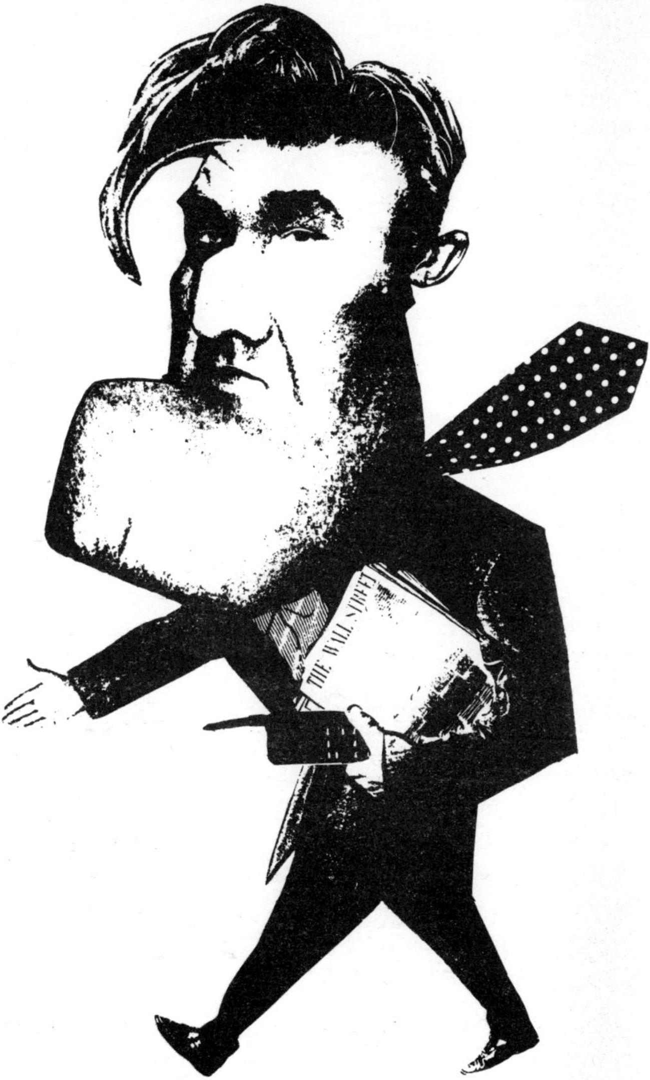
THE WALL STREET

The Lovebirds

There is an old saying that you should never shit in your own backyard (presumably shitting in someone else's backyard is acceptable), but five minutes in the company of the Lovebirds and you'll be amending your CD-ROM of CRAP OLD SAYINGS to you should never *date* in your own backyard.

Gurgling to each other in baby talk that a baby would find embarrassing, wearing identical jumpers and having similar interests, this couple are so nice that they make Mother Teresa look like Virginia Bottomley.

They will start out as sad, lonely individuals feeling the first touches of SOS, but then their eyes will meet across the IT department. It will be love at first sight. From that second on they will spend every second in each other's company, believing that an SOS symptom shared is an SOS symptom halved (time to update that CRAP SAYINGS disk again).

What they won't realize is that an SOS symptom shared is an SOS symptom doubled, the sickness causing their minds to merge into one single, four-lobed cuddly toy.

The office will carry on around them, their attentions wholly consumed with the planning of THE WEDDING, an event scheduled for some-time in the next decade.

Every nuance of the occasion, from the doilies embossed with the happy couple's initials and birth-signs, to the exact running order of the honeymoon night (accompanied by much snogging and smothered giggling) will be discussed *ad nauseam*, Pronuptia catalogues and copies of *Bride* magazine mounting in the staff room like glossy termite hills.

Visual Manifestations of SOS

Starry eyed and joined at the lips, they will both make extravagant hand gestures in order to bring attention to their his 'n' hers Forever Friends wristwatches.

Habitat

They will have matching desks on either side of the office, each displaying gilt-framed photographs of the other's parents.

Diet
Scrummy li'l nibbles that they'll feed to each other.

Effect the Electronic Age will have on Relieving Their SOS
They will both be disappointed that the lack of a certain key prevents them from sending each other I ♥ YOU messages by E-mail.

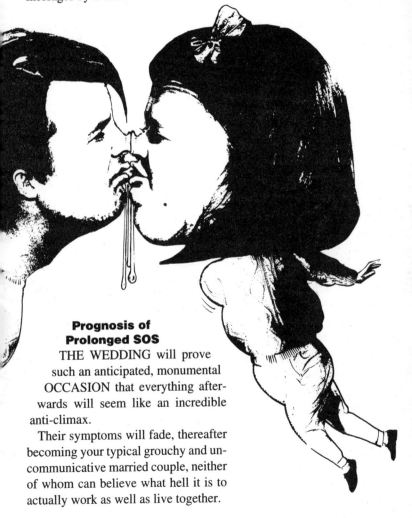

Prognosis of Prolonged SOS
THE WEDDING will prove such an anticipated, monumental OCCASION that everything afterwards will seem like an incredible anti-climax.

Their symptoms will fade, thereafter becoming your typical grouchy and uncommunicative married couple, neither of whom can believe what hell it is to actually work as well as live together.

Mister Whitebread

A bitter little man, this staff member believes that he has a God-given right to respect and career opportunities, being MALE and WHITE. He struggles to come to terms with the nineties multi-racial, sexually equal, politically correct office environment, but privately keeps a long list of racial groups, social types and sexual orientations that he blames for his own dismal life, world problems, the decline in the public transport system, etc.

He never actually spouts his racist views – he's too spineless to do that. He just drops little comments to female colleagues like, 'What's the point of you going after promotion when you'll be making babies in a couple of years?' and then blames the 'time of the month' when they react badly to this.

Every now and then he exposes his true feelings by regurgitating Bernard Manning 'jokes' and then telling anyone who takes offence that they should 'get a sense of humour'.

Mister Whitebread will run into great personal conflict when a culturally diverse lesbian is appointed his boss, grudgingly following her every command whilst walking with buttocks so preternaturally clenched that he shits diamonds.

His prejudices will be continually compounded by SOS, driving him so far from the pit stop of sanity that BNP policy will actually make sense to him.

Visual Characteristics

In his SOS-addled mind: a strapping Englishman, the proud product of a tiny country that built an empire!

In reality: a pitiful inadequate with brown shoes and a scalp like flaky pastry.

Favourite Movie

A toss-up between *Falling Down* and *Zulu*.

Diet

'None of your foreign muck – just good British grub!'

(For overseas readers that's fatty meat served with lashings of carbohydrates in a reservoir of grease.)

Prognosis of Prolonged SOS

He'll explode after being appointed company Equal Opportunities Officer.

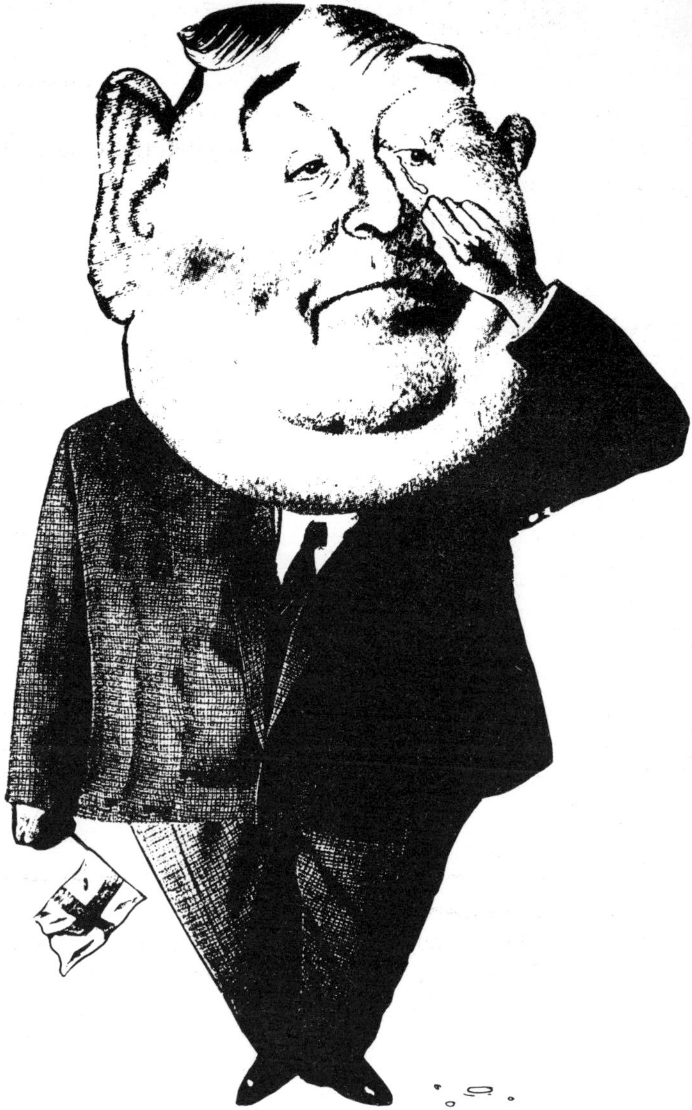

The Militant Unionist

Once a common sight in every office in the country, this affectation has now become one of the rarest forms of SOS, only found in isolated Northern areas and completely extinct anywhere south of Birmingham.

The condition arises from the belief that management is responsible for every discomfort/misfortune that occurs to employees. This may be a correct assumption but it is one which most workers choose to ignore as long as they get their monthly wageslip and twenty-five days holiday a year. The general lack of interest in the office will cause the Militant Unionist to rally even harder, and as SOS takes hold, so the pettiness of the targets increases. The '95 HORRIBLE ORANGE PAINTWORK NO MORE! campaign will be followed by the ADJUSTABLE FOOTRESTS ARE A RIGHT, NOT A PRIVILEGE! initiative, all the time ignoring the fact that there hasn't been a wage rise in three years, staff breaks have been cut and there still aren't proper childcare facilities for working mums.

As membership dwindles, this character will have to stand for an increasing number of positions on the committee; President, vice-president, chair, vice-chair, treasurer, organizer, scrutineer, and person who initiates a chorus of mumbling after every contentious notion is floored. Eventually the Union has to change its motto from 'One out, all out' to 'All out, one out'.

Visual Manifestations of SOS

An angry scowl and a Broad Left t-shirt, often complemented by an Issue of the Day button badge.

Habitat

The Union Room, a cramped little office where hyperbolic leaflets are put together in BOLD TYPE, with titles such as *The Truth About . . .* and *More Management Lies About . . .*

Catchphrase

A variety of football analogies including, 'The management have moved the goalposts', and 'The management have played a long shot down the flanks, killing the ball dead before finally showing our strikers what good finishing is'.

Prognosis of Prolonged SOS

The symptoms will disappear
completely the moment the
Unionist either receives
promotion or obtains
a mortgage.

The Design Artiste

A tortured, misunderstood soul who works in the design department, the Design Artiste will make everybody's life a misery – despairing over the philistinism of clients and treating every paste-up job like it's the Sistine Chapel.

He will finish his foundation course and have great plans to shake the art world with his bold painting techniques. Unfortunately, a cosy middle class upbringing will ill prepare him for the required starvation period and so he'll land a design job to keep him in 'brushes' (hey, even the Impressionists had patrons!). SOS will soon cause him to lose all perspective of what he's actually paid to do, ignoring clients' briefs and producing huge mixed-media installations when all is required is an A5 flyer announcing a change of address.

Every time one of his cutting-edge grid constructions based on chaos theory or obscure typefaces derived from the primitive script of Aboriginal cave painters is rejected, so his madness increases, eventually reaching a level of insanity on a level with Gilbert and George.

He knows that his true genius will be recognized after his death, just like Van Gogh. Whether his SUMMER REDUCTIONS ON MANY LINES poster will have the same lasting quality as *Sunflowers* remains to be seen.

Visual Manifestations of SOS

Tortured brow, sensitive features and a dress sense that screams DESIGNER. Never seen without his FONT FONT/HEADLINE typeface catalogues tucked under his arm.

Habitat

Moping around the design area chain smoking or at home watching European films/browsing through American 'new art' magazines, searching for inspiration (i.e. something to rip off – the curse of post modernism).

Greatest Achievement

Winning the award for Most Unreadable Typeface of the Year in *Emigré*, the magazine of unreadable typefaces.

Effect the Electronic Age will have on Relieving His SOS

Will actually increase his symptoms. The advent of the Apple Mac has cut the actual time of *doing* a design job down to ten minutes before the deadline. This means he has more time to despair over insensitive 'slave drivers'.

Prognosis of Prolonged SOS

Just as his hero Van Gogh cut off his ear in a peak of nervous energy, in a fit of pique he will snip off the right earpiece of his Sony Discman.

Mister Negative

The loss of a beloved stick insect, a narrow escape from death in the bowels of an NHS trust hospital (or Donor Farm for BUPA Patients as they're often known) and the landing of an office job will give Mister Negative a fatalistic attitude bordering on . . . well, the *fatal*.

It's then obvious that SOS will intensify this feeling as, let's face it, *work is depressing*.

He'll soon be despairing about everything – the effect of the photocopier on the ozone layer, his pitiful wages, wondering if his VDU is helping to swell the brain tumor that he suspects he has. When he's got nothing to be depressed about he turns to the newspapers and soon has something awful to share with the world.

Suggest that he tries professional counselling or a course of antidepressants to lift him out of this sobbing-at-his-desk, composing-suicide-notes-and-storing-them-on-file condition and he'll reel off a list of specialists he's currently consulting and uppers he's been prescribed, just to get him up to *this* level.

He will not be a popular man in the office. The general feeling is that his suicide would put everyone out of their misery.

Visual Characteristics of SOS
Moody and distant, he will take to wearing smiley badges upside down.

Habitat
Mooches around the corridors, disseminating rumours of company collapse and spreading bad cheer, the Cassandra of the valium generation.

Diet
Always eats well as it's probably his last meal . . .

Motto
Life is shit and then you die. Which is also shit.

Favourite Music
Lou Reed's *Magic and Loss* album. He considers Morrisey/Leonard Cohen to be distracting party music.

Prognosis of Prolonged SOS

His symptoms will actually fade after receiving an office collection, the gift causing him to realize that people *do* care.

What he won't know is that the whip round was to buy him a length of rope.

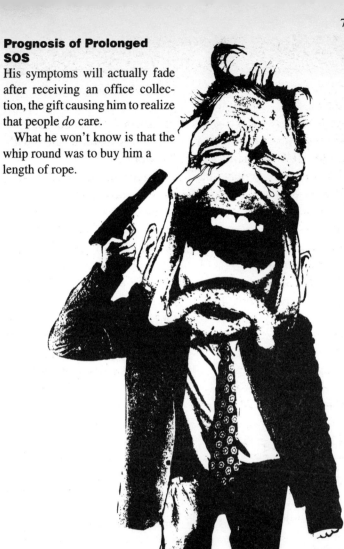

The Miserly Stationery Officer

Like the Middle Management Sadist, the Miserly Stationery Officer develops full-blown SOS the minute he is handed any degree of responsibility.

The ordering and distribution of stores dominates his waking hours, the counting of A4 plastic ring binders jumping over a fence being the only way he can get to sleep at night. As his symptoms worsen, so his desire to hoard grows, making it virtually impossible to get any stationery out of him whatsoever. He will stockpile whole Amazonian rain forests in his tiny storeroom.

He perfects paperwork as a stalling technique. Request a pen and he'll roll his eyes and insist that you fill in the necessary form stating pen colour, line width, tip type (ball or felt?), serial number, batch requisition code and delivery location, all completed in triplicate.

Another facet of his sickness will be his one-man crusade against stationery robbery, believing that staff helping themselves to the odd Biro is just the Tipp-Ex of the iceberg. Countermeasures implemented will include internal body searches at home time and computer tagging of office hole punchers.

Visual Manifestations of SOS

Identifiable by his rubber paper-thimble and the mighty set of storeroom keys swinging dangerously from his belt, he will wear his paper-cuts as a badge of office.

Habitat

An 8' x 8' stationery cupboard, in which he huddles amongst the paper like a gigantic nesting hamster.

Favourite Holiday Destination

Manila, his spiritual birthplace.

Diet

Normal healthy eating practices except rice paper, which he considers cannibalism.

Office Affiliations

Refuses to (Basildon) bond with any staff members ('They're all after my paper. My beautiful paper!') but will spend hours courting stationery reps in a strange language consisting of phrases such as 80 gsm pastel and Rexel 10-line 101 staples.

Effect the Electronic Office has on Relieving His SOS

See Prognosis (below).

Prognosis of Long-term SOS

Will eventually lose his position when the office becomes exclusively digital; is subsequently shot by police marksmen after a two-month siege to remove him from the store-room.

Eddie – An SOS Case History

Okay, you're now aware of SOS – its devastating effects, and various manifestations, but it all still seems distant . . . theoretical, not something that could happen to *real* people.

What follows is a true life case history tracing Eddie's development as he is torn from the comforts of unemployment and thrown into a dark and unfamiliar world.

Only his name has been changed (it's really Peter Squires) to protect the feelings of his friends and family (would you want it publicly known that you associate with an administrative assistant?)

Eddie had been unemployed since leaving university, attending the odd interview just to keep the DHSS happy. One Thursday in August, Eddie had just got up (recovering from watching late night television) and was about to make some toast and settle down in front of *Take the High Road* when the telephone rang.

It was the personnel manager of a local manufacturers where Eddie had attended an interview the previous week for a position in the administration department. The voice on the other end of the line was to the point: 'You've got the job,' it said.

No friendly preamble.

No cushioning the blow.

Eddie was shattered.

He started work the following Monday and soon entered the downward spiral that is SOS.

He began to live for payday (finding the more he earned, the more he owed), quoted from lame comedy shows that had been on the television the night before and drank himself senseless on Friday night before wandering around the streets shouting, 'Wey-oh! Wey-oh! Wey-oh!'

Obviously, he fitted right into the office and as he became more settled he was handed a little responsibility, the head of the administration department allowing him to make his own decisions.

It is at this point that he discovered he held the power of obfuscation over the whole office, vetoing any new projects put before him and backing up his decision with technical terms, reams of printed evidence and tons of 100% homegrown bullshit. The effect of this also helped him further down the line – no new projects meant nothing for him to administrate later on.

Of course he obviously had to green light some projects but only after the developers had gone back to the drawing board, jumped through hoops and kissed his butt.

The logjam known as Eddie became a source of great frustration, Eddie's superiors backing his opinions as their workloads also lightened considerably as he blocked still more.

Eddie is still in his position and is currently training an assistant to obfuscate developments down to a trickle which he will then be able to obfuscate at a more senior level . . .

SOS Workstation

Employee's desk environments tend to reflect the particular strain of SOS that they are suffering. Thus, the Habitual Post Graduate's desk is laden with rag mags, back issues of the *NME* and a bust of Tutankhamun, whilst the Lager Loutette's features an array of empty Pils bottles, a KISS ME QUICK, SHAG ME SLOWLY hat and a BARE-ARSED BOYS ON BIKES calendar. There are, however, certain staples (meaning main elements and not the U-shaped clips used for ripping out fingernails and breaking scissors) that can be found in every long-term employee's workstation.

VDU, the electronic heart of the workstation. Keyboard key spaces containing crumbs of chocolate, cake, sandwiches, spicy beanburgers, etc. Serves as a tasty smörgasbord on those long, hungry winter afternoons.

Phone with memory store containing the numbers of friends and family, all the food delivery services in a five-mile radius, various competition hotlines and the Samaritans.

Calculator with Flexi-time Calculation Function for working out time down to the nano-second.

Wastepaper bin full of empty plastic coffee cups (drinking that much caffeine isn't advisable but it takes three minutes and forty seconds to walk to the coffee machine and back and that's three minutes and forty seconds away from the workstation).

Chair positioned at an uncomfortable, unergonomic angle to encourage spinal difficulties and necessitate time off work to visit the osteopath.

Picture of a loved one be it wife, husband, child, pet or chocolate eclair.

77

erky, humorous sign, obviously conceived y someone who's never worked in an office. f it was truly reflective of the zeitgeist it would read I'M SO F**KING BORED I WANT TO DIE.

Coffee mug bearing the legend WORLD'S GREATEST MUG (DAD/MUM/HUSBAND/ GOLFER/CLERICAL OFFICER), a gift from a loving relative. Apparently, the script is not meant to be ironic.

Calendar counting down the days to retirement.

Carrier bag full of stolen stationery (the single advantage of working in an office).

A blank piece of paper on the desk top or carrying around the office in order to ook busy.

Top drawer stash consisting of a walnut whip (for those bad days), a rapid fire Prozac dispenser (for those really bad days) and a noose (for Monday mornings).

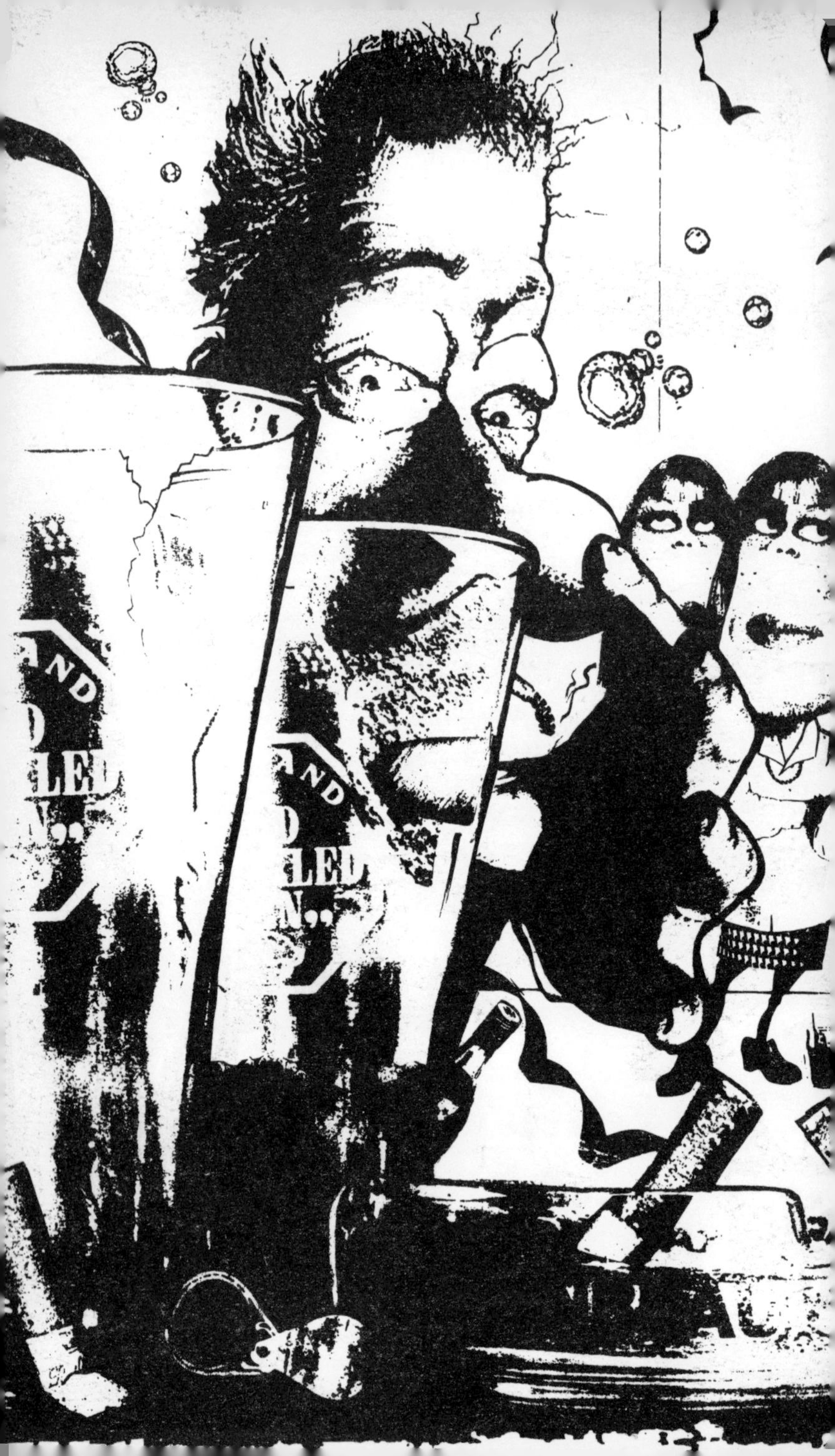

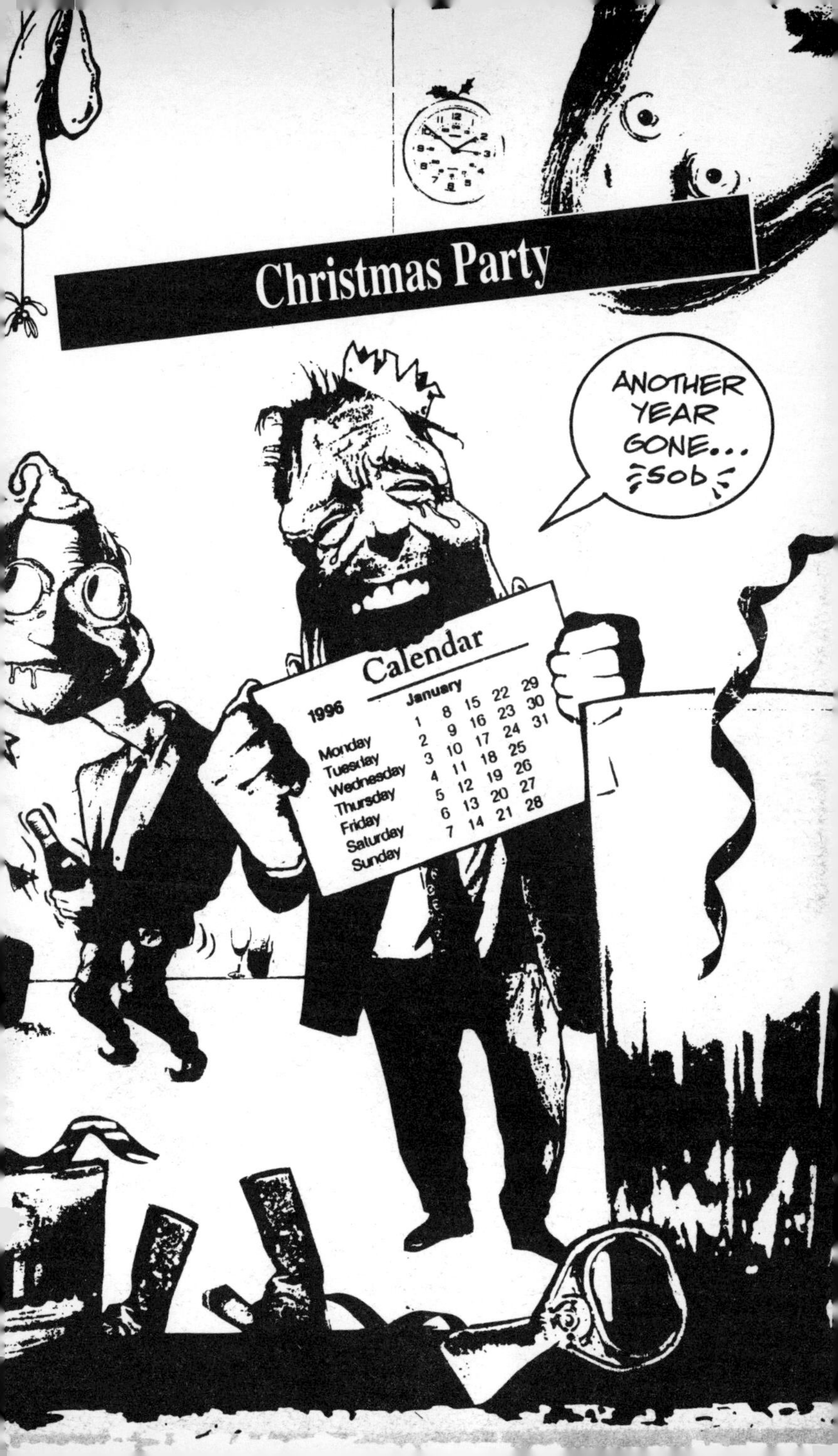
Christmas Party
ANOTHER YEAR GONE... sob
Calendar
1996 January
Monday 1 8 15 22 29
Tuesday 2 9 16 23 30
Wednesday 3 10 17 24 31
Thursday 4 11 18 25
Friday 5 12 19 26
Saturday 6 13 20 27
Sunday 7 14 21 28

Christmas Party

Once a release from the pressures of the year – a real bacchanalian orgy of drink, food, sex and mouth-manipulated party blowers – the Christmas Party has gradually been eroded by the collective effects of safe sex, political correctness, drink-driving anxieties, redundancy paranoia and, most of all, the realization that another precious year of your life spent in the office is *nothing* to celebrate.

So the Christmas Party is now regarded as something of a chore. You show your face, make smalltalk ('I haven't bought any presents and there's only three shopping seconds till Christmas!'), drink one can of cheap lager (if supplied by management this will be well past its DRINK UNTIL INSENSIBLE BY date), avoid being kissed by your tuna breathed/ cold-sore-encrusted boss/secretary, make your excuses and then go and celebrate the yuletide season with people you actually like.

A seasonal atmosphere will be attempted by the cursory draping of tinsel around the office fittings, sad strands of silvery string that are resurrected every year and which were probably originally bought at the Bethlehem branch of Poundstretcher by one of the shepherds. These will be left hanging in the office until June when one of the SOS sufferers will blame his bad luck on their post-twelfth-night presence.

Remember, 'tis the season to be jolly, so you're probably better off skipping the party completely, staying at home and watching It's a Wonderful Life.

Useful Objects to Help You Survive the Christmas Party

- A handheld mistletoe defoliant.
- A well-rehearsed and vaguely plausible excuse for an early exit.
- A camcorder to capture any hint of impropriety, to ensure you'll never have to make the tea for the rest of the year.
- A box of ethnic charity Christmas cards, 'handmade by dyslexic Peruvian crossword compilers'. These can be knocked up the night before the party using cheap crêpe paper and illustrated with your feet, and provide a cheap and sensitivity-confirming seasonal greeting.
- A certificate of Jewish conversion, a surefire excuse for non-attendance.

Bomb Scare

This has replaced the Christmas Party as the one enjoyable office social event, when staff members can bond together in an air of mutual respect and general bonhomie.

Maybe this is because it's a chance, unexpected opportunity to escape from the workplace or maybe it's the sense of danger, but something about standing in the road preventing the flow of traffic rekindles the Dunkirk spirit. Indeed, the bomb scare is only some bunting and a couple of sausage rolls away from a street party. If terrorists and crank callers could give you a couple of weeks' notice then a light cabaret/outside caterers could be arranged.

Office staff everywhere have reacted badly to the recent ceasing of hostilities by the IRA. The subsequent lack of mainland bomb scares is thought to be behind the rush of phone calls to the Iranian embassy claiming that Salman Rushdie is the janitor in various city offices.

NB Don't appear to enjoy yourself too much during bomb scares, as you may be collared by over-enthusiastic members of the emergency forces under the auspices of the Criminal Justice Act.

Useful Objects to Help You Survive the Bomb Scare

• A warm coat to help you withstand the British winter weather.

• A warm coat to help you withstand the British summer weather.

• A padded envelope containing an alarm clock to be left in the office where it can easily be found by army bomb disposal experts. Should ensure you don't get back in the office for the rest of the afternoon.

• A receipt for the Rolex you had in your desk drawer and the phone number of the claims department of your employer's insurance company, just in case there really is a bomb.

• A wheelchair to ensure you get a fireman's lift down all those stairs.

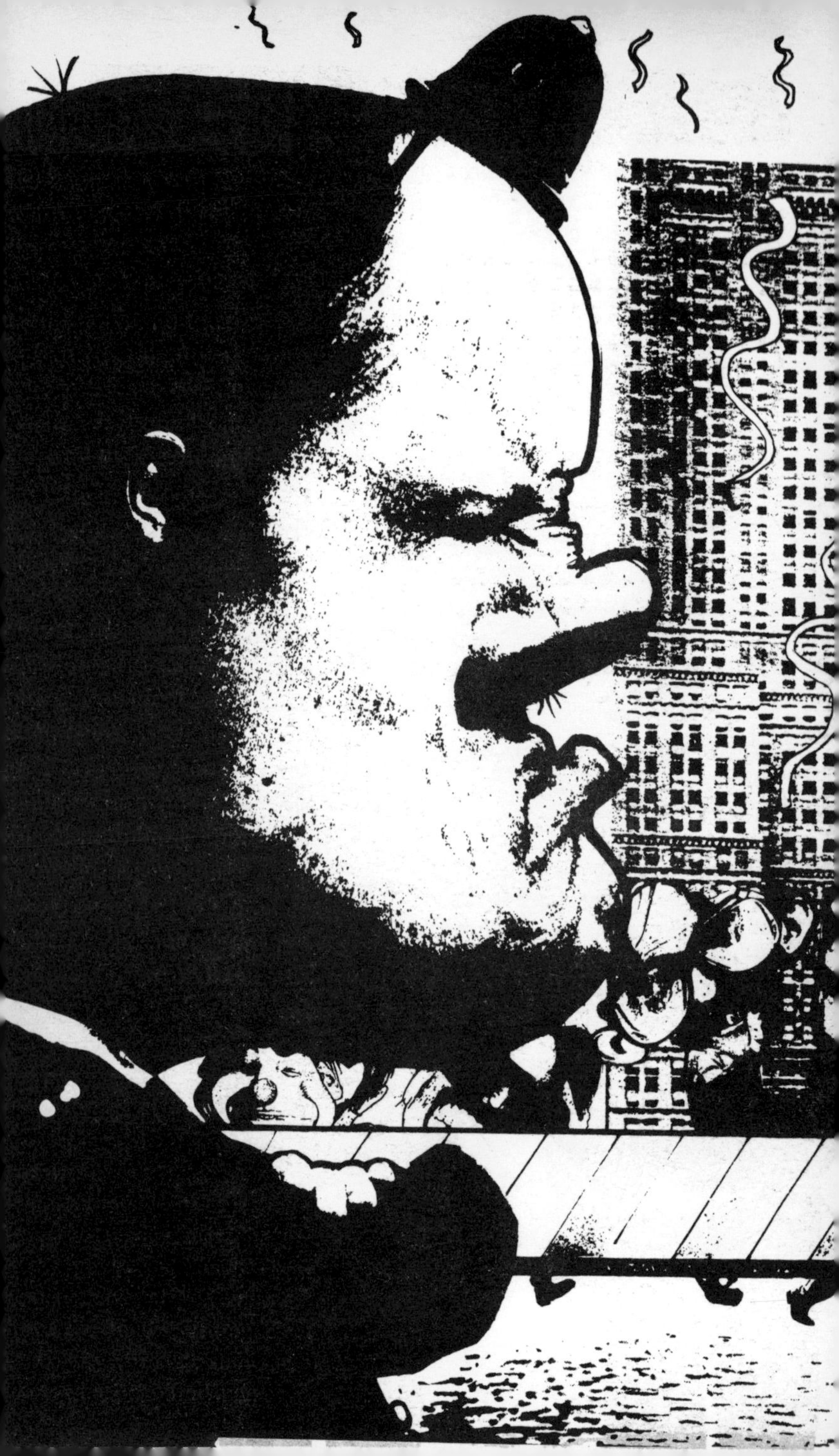

Bomb Scare

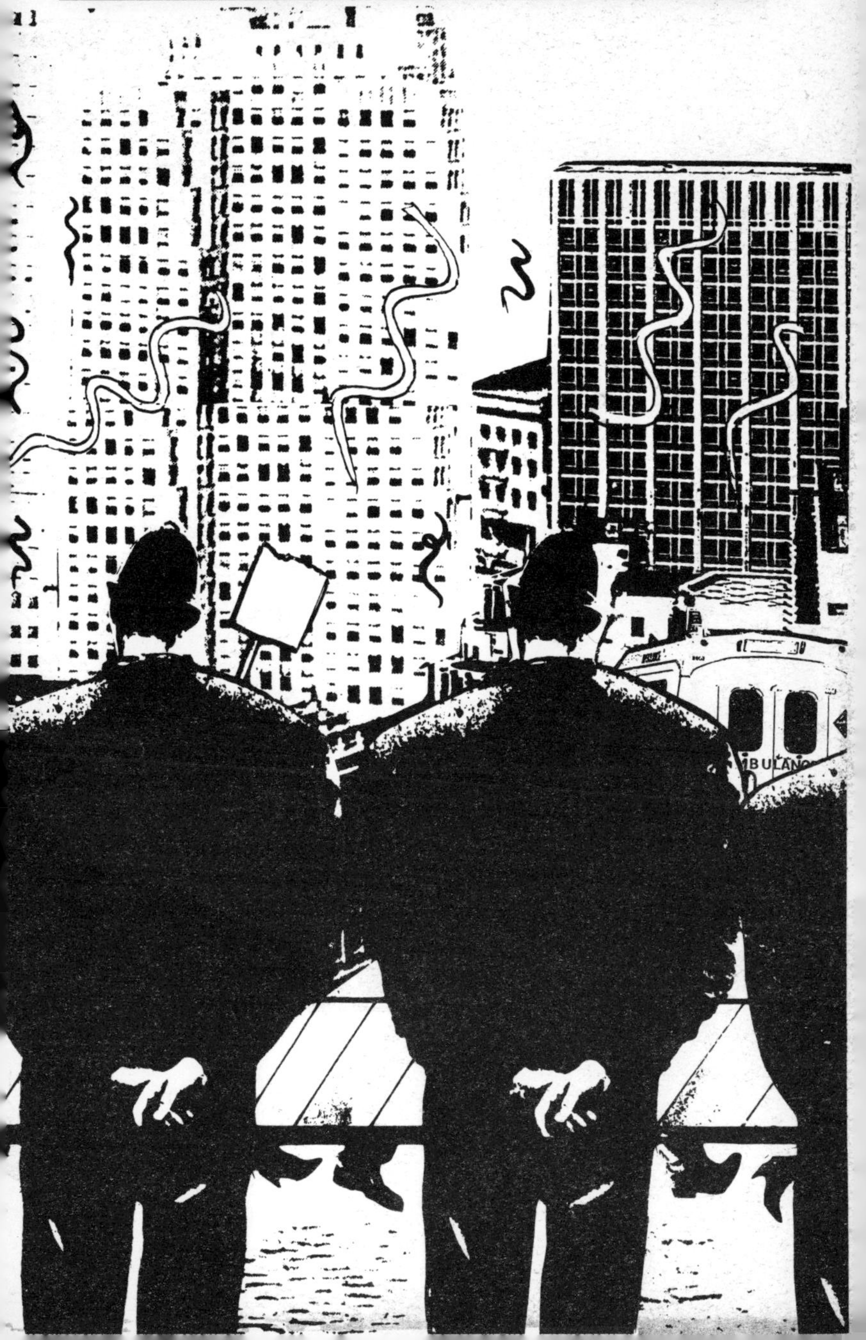

The Annual General Meeting

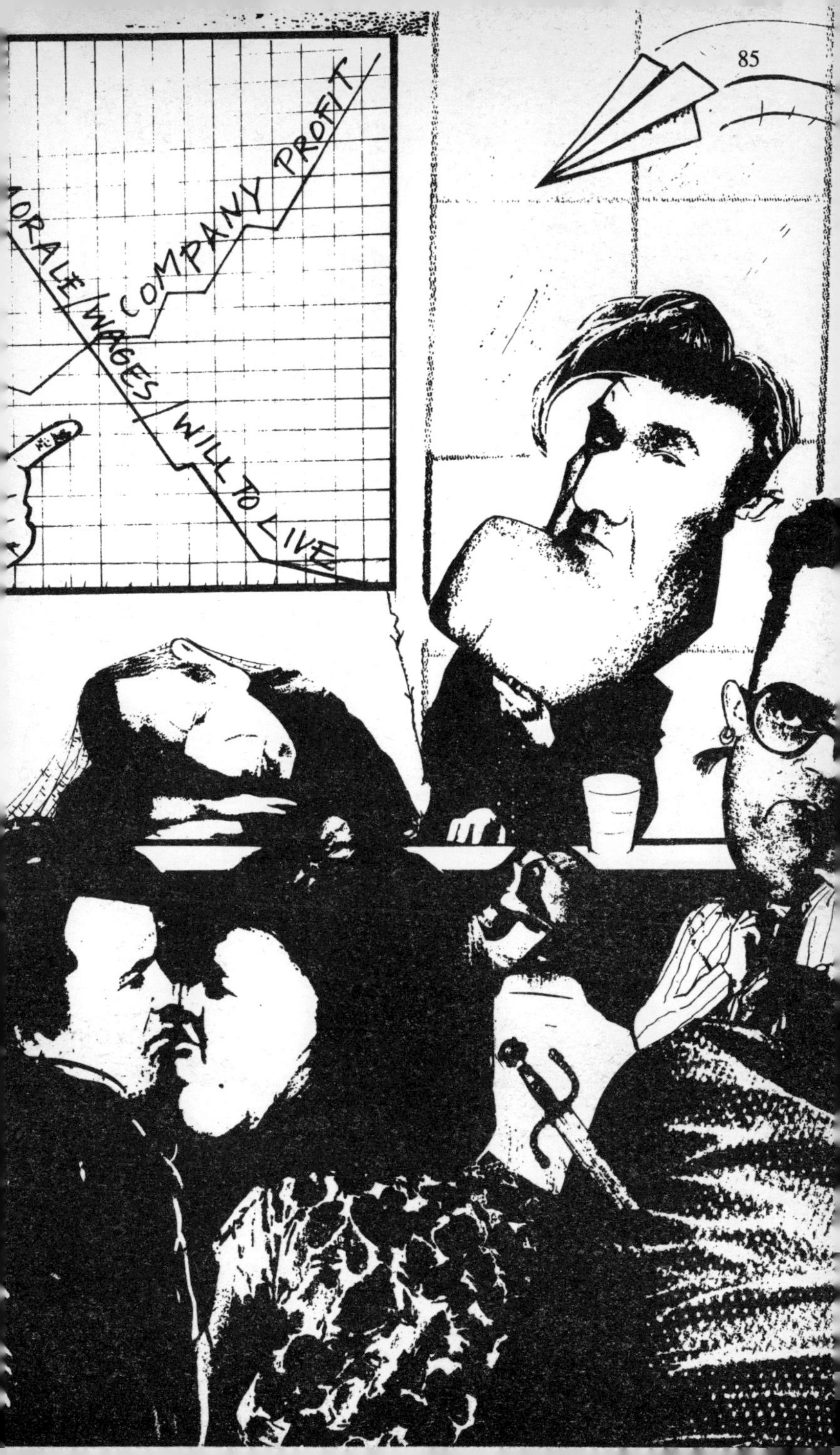
COMPANY PROFIT
MORALE
WAGES
WILL TO LIVE

The Annual General Meeting

Even though it provides much-needed time away from the work-day grind, the Annual General Meeting (AGM) is the most dreaded day of the year.

Staff are shepherded together and addressed, at length, by senior management on the company's previous twelve months, its current standing and short/long-term future. The formality of the affair brings to mind a school assembly, an event (like a wedding service) where you are gripped by a super-natural urge to jump on a table and yell 'bollocks' loudly at the most inopportune moment.

The thrust of the meeting is that senior management have spent half a million pounds and six months on a report on how to make the company more cost and time efficient.

Their findings are, *either*:

The company is barely managing during troubled times and for it to even survive staff are going to have to work harder, accept a cut in pay, have the heating turned off during winter months, sacrifice their coffee breaks and only flush work toilets once every fifteen functions. *Or*:

The company's performance is exceptional during troubled times, but for it to stay effective staff are going to have to work harder, accept a cut in pay, have the heating turned off, etc.

If the company had somehow stumbled across a combined cure for cancer, male pattern baldness and panda impotence, senior management would still trot out the same spiel.

To make matters worse, this BS is delivered in the most arse-numbing doggerel imaginable, putting Iraq's Elite Republican Bore Corps (a crack troupe trained in tedium using such UN-prohibited techniques as trainspotting numeracy, estate agency smalltalk and Trekkie chitchat) to shame.

Photocopies of interim figures, prospected shortfalls and company targets will be distributed and mean as much as a braille colour chart to a blind man with no arms.

As you are lulled to sleep by the aural equivalent of the Chinese water torture, remember this – the company shareholders are there by choice!

Useful Things To Help You Not Survive The AGM

- A hollowed tooth containing a life-threatening dose of cyanide.
- A seat near an open window.
- A fatal heart defect that is triggered by the repeated use of the phrase PROSPECTIVE LONG TERM PROFIT MARGIN.

NB The recent deaths by nerve gas on the Tokyo underground just happened to take place the morning of the Tokisha Corporation's annual general meeting. Coincidence?

Glossary

Apple Mac

A state-of-the-art computer hardware/software package whose presence can be found in every office but yours. If you do happen to have one, the queue will be horrendous. Still the interdepartmental memos have never looked so good.

Annual Leave

25 days a year holiday time that employees are given to wait in for the gas man, the cable installer . . .

Administration

A department designed to ensure the flow of paperwork, wages, annual leave, etc., is as moribund as possible without being entirely stationary. Treat all admin workers cordially or they will make your life HELL.

Bank Holiday

A day set aside for the British working public to watch old James Bond films.

Barcode

Never spit at the barman and always pay for your round.

Coffee Machine

Humorous name given to poisonous chemical sludge dispenser found in company staff rooms. Thoughtful suppliers will label 'leaded' and 'unleaded'.

Childcare

The concept that companies actually value/support their female employees and so provide onsite crèche facilities to allow them a worry-free return to work. Unknown in the United Kingdom.

Company Car

These are post-boom years, this is the bus.

Fax

A device used for turning a sheet of A4 double-spaced typing into something resembling a three-month maternity scan, and shredding it to anywhere in the world via a telephone.

Flexi-Time

A system which enables employees to work hours to suit themselves. This means that the whole office has extended lunchbreaks for the first three weeks of the month and then has to work a 75-hour week in the final quarter to make up the time.

ID Pass

A laminated pass issued in many offices for security purposes. This will bear your name, name of your department and a photo of an international terrorist surprised by a carrot.

NB Only dorks wear them on their belts.

Job Sharing

A way of working where a number of people share the work and responsibility of one full-time job. Also known as the Civil Service.

Industrial Relations

This once meant the relationship between staff and management. As this no longer exists, it is now only used to describe in-laws employed in light engineering.

Keyboard

Sometimes used on boring mornings to make the cursor whip around the screen in an approximation of Pacman. More often employed as a snack dispenser and storage container for Kirby grips, nailfiles and paperclips.

Lunchbreak

Free time used by staff to shop, go to the pub, queue up in banks, etc., in fact everything but eat lunch, which is consumed at the desk around 11.30 a.m.

Mailroom

A gallery space for an impressive assortment of tabloid-clipped Elizabeth Hurley/Pamela Anderson/Naomi Campbell pictures, marred only by mountains of undelivered packages, mailbags, etc.

Management

Usually referred to as 'Them'. Like the Olympian gods of Ancient times, they will be both omnipotent and utterly invisible, unless they choose to mess around with the lower mortals by manifesting themselves as, say, a shower of memos or a plague of downsizing initiatives. They will dispense directives from on high, in a suitably cavalier and seemingly arbitrary manner, blissfully unaware of the chaos that they will cause at grass-root level. In fact every move they make is carefully planned – to keep them in the gravy, and everyone else in the lumpy custard. Any 'decisions' needed to be made by the management are usually avoided for as long as possible, and are often finally resolved by a lower executive casting the runes, or slaughtering a sacrificial trainee.

Memo

A note to inform the staff of the painfully bloody obvious.

Maternity Leave

Almost worth the morning sickness, water retention and varicose veins. But not quite. Still jealously guarded as a female preserve – but who knows how desperate men will get for time off come the millennium?

Overtime

Work above and beyond an employee's allocated worktime, done

Pension

Money set aside to provide cat food in your old age.

Photocopier

A piece of surrealistic sculpture to be found in most offices. Of no known use. Often enigmatically titled ENGINEER CALLED.

Post-It

Adhesive label used to place ironic 'comedy' signs on other employees' backs.

in order to keep their increased workload to a barely acceptable level. Usually unpaid.

Part-Time

Those employees who only work a percentage of the week. Sorry, that should read those employees who *officially* only work a percentage of the week.

Personnel Department

An empty office with '*Personnel Department*' written on the door.

Sideways/Internal Promotion

More responsibility, same pay. Now the only 'promotion' ever offered. Also known as 'organic growth'.

Sick Leave

Additional days off used to wait in for the gas man, the cable installer, etc.

Shredder

See Fax.

Telephonist

Employee who mans/womans the office switchboard and whose job it is to treat callers in an abrupt manner before casting them into telephonic limbo where they are bombarded with a Euro-pop arrangement of 'Greensleeves' before being cut off altogether.

First line of defence against irate clients.

Temp

A troubling metamorph, who spends weeks learning the rudiments of the task she's been assigned and who then adopts a new form that *looks* the same but totally lacks her new-found abilities. And so the cycle continues . . .

Tipp-Ex

An executive toy applied like nail polish during long boring phone calls.

Training

(Archaic.) Always carried out 'in house'. The placement of a completely unqualified member of staff into a job of which they know nothing, for the same pay. Also known as 'Sink or Swim'.

VDU

A rectangular monitor used to display old Post-it notes.

Voice Mail

A device whereby if a caller manages to make it through previous hurdles (see Telephonist), he/she is then met by a barely audible message, blasted by a long 'beep' that lasts twenty minutes. Finally they are given 4.5 seconds to leave a message.

Second line of defence against irate clients/spouses.

Wage Slip

Much anticipated but always disappointing printed record of money theoretically owned for a nanosecond before it disappears into the void of your overdraft.

Word-Processor

A computer system that ensures your work is professionally and efficiently typed before it is professionally and efficiently eradicated from the entire disk.

Afterword

So you've read the contents of this book and suddenly your future looks bleak. All that lies ahead is increasing dementia caused by a series of days which are all carbon copies of the one you've just endured.

Well don't be too depressed!

THINK POSITIVE. The only way to survive your working life is to stop and smell the daisies. Relegate your job to the sidelines of your existence and enjoy everything else that life has to offer, like:

A Supportive Partner

There is nothing more comfortable than having a loving companion to talk at you as you drift into a coma in the armchair each work night. Of course, if *they* don't have a paid job then they'll probably be suffering from Sick House Syndrome (a whole other can of worms) and you can expect to be nagged to death or eviscerated with a carving knife if you walk through the door after forgetting to pick up some milk/a newspaper/hard spirits.

A Family

Children are a constant source of fun and distraction. Take them into your office every now and then as a warning of what will happen if they don't do well at school.

Holidays

Two weeks in the sun will lift your soul, clear your mind and straighten out your servitor's stoop.

Don't even *think* about the office when you're lounging in the sun – pretend you haven't even got a job to return to. Make believe you're a young offender sent abroad on tax payer's money as a short sharp shock treatment, the only thing beyond these two blissful weeks being a safari holiday in Africa for a further offence.

A Social Life

Get singing-with-strangers drunk at the weekends. It's a laugh. And after all, you then have five days to slouch around and do nothing whilst you recover from your hangover, ready for the next weekend.

Television

The great healer. Allow four hours of soap operas, game shows and weak sitcoms to wash over you every night and let your troubles seep away.

This also has a bi-product in that it gives you something stimulating to talk about next day at work – 'Did you think Cilla was wearing that suit for a bet?'

Remember, however bad work gets, things could be worse. At least it's not the eighties when you had to toil at a crappy job and pretend it was all you ever lived for . . .

About the creators of this book

A bombastic combination of piss artist and office plant, Carl Flint enjoys baking cakes and resting, has never had a proper job in his life and has especially never worked for *NME, Select, Readers Digest* or *Sonic the Comic*. He currently awaits being abducted by aliens.

/co-creator *the Skel-* to many cluding *, Tank load,* ently sey.